Racecar Technology Level One

Learning All About The Parts and Pieces Of A Race Car
An Entry Level Racers Guide To Race Cars

Bob Bolles

CRD Publishing

CRD Publications

Published by Chassis Research and Development Corporation, aka/Chassis R&D

P.O. Box 730542, Ormond Beach, FL 32173-730542. USA

Email: chassisrd@aol.com

Race Car Technology – Level One

Copyright © 2019 by Chassis Research and Development Corporation

All photos by Bob Bolles unless otherwise noted.

All rights reserved. No part of this book/publication may be reproduced, scanned, or distributed in any printed or electronic form without permission. Please do not participate in or encourage piracy of copyrighted materials in violation of the author's rights. Purchase only authorized editions.

CRD Publications is a trademark of Chassis Research and Dev. Corp.

PRINTING HISTORY

First CRD edition / June 2019

ISBN: 978-1-7324884-3-4

CREATED/PRINTED IN THE UNITED STATES OF AMERICA

Cover design by: LS Bulow, Graphic Designer

NOTICE: The information in this book is true and complete to the best of our knowledge. All recommendations on parts and procedures are made without any guarantees on the part of the author or the publisher. Author and publisher disclaim any and all liability incurred in connection with the use of this information. We recognize that some words, parts names, model names, and designations mentioned in this book are the property of the trademark holder and are used for identification purposes only. This is not an official publication.

Racecar Technology – Level One

Table Of Contents – RT Level One

Lesson 1 - Goals of Race Car Technology Level One	3
Lesson 2 – Why A Race Car Turns	7
Lesson 3 – Camber	11
Lesson 4 – Caster	15
Lesson 5 - Spindle Design	19
Lesson 6 - Control Arms	25
Lesson 7 – Anti Pro Dive	29
Lesson 8- Steering Systems	33
Lesson 9- Bump Steer	37
Lesson 10- Ackermann	41
Lesson 11- Alignment	45
Lesson 12– Sway/Anti Roll Bars	49
Lesson 13 – Front Geometry and Roll Center / Moment Center	53
Lesson 14- Types of Rear Suspensions	59
Lesson 15- Rear Roll Center	65
Lesson 16- Alignment and Toe	69
Lesson 17- Rear Steer/Roll Steer/Bump Steer	73
Lesson 18- Anti-Squat	77
Lesson 19- Driveline Alignment	81
Lesson 20 - Types of Springs & How To Rate A Spring	85
Lesson 21 - Motion Ratio & Wheel Rate	93
Lesson 22 - Spring Base Principles & Choosing Spring Rates	97
Lesson 23 - What They Do and Do Not Do	101
Lesson 24 - How A Shock Works	105
Lesson 25 - Different Types of Shocks	111
Lesson 26 - How To Control A Spring	117
Lesson 27 - Motion Ratio Affects Shock Rate	121
Lesson 28 - Tire Temperatures & Tire Pressures	125
Lesson 29 - Tire Stagger	133
Lesson 30 - How Racing Brakes Work	139
Lesson 31 - Summation for Level One	145

INTRODUCTION

Around 1992 I decided to change careers and become a race car engineer. My early experience in racing as a kid was spending countless hours at Daytona International Speedway in the pits, in the stands and around the mechanics, drivers and owners listening and hearing about how the cars handled. I was always fascinated by the design and setup of race cars, be they stock cars, road racing cars, formula or weekend SCCA cars.

I attended many years of races at New Smyrna Speedway and Barberville, now known as Volusia Speedway Park. I was interested in the "race", but always fascinated by the way the cars handled and how all of that was accomplished. I watched Dick Trickle prepare his car in 1975 at a friend's garage in preparation for the Speed Weeks show at New Smyrna and thought, I would like to be able to do that.

I am an engineer by education, degree and by nature and I knew someday I would have to get involved with racing. When that day came, I threw myself into the task of learning and inventing with more energy and determination than at any point in my life with anything I had ever done. It was a passion combined with a purpose. I was determined to find the truth about chassis dynamics and race car setup.

My work, and indeed my racing business, was born out of frustration and failure in trying to find really helpful information that I could use to set up a racecar. So, I set out on a journey that followed in no one's footsteps. Instead, I used one of my greatest personal assets, a profound and acutely developed ability to apply a common sense approach to problem solving. That is exactly what you will find in this book, a common sense approach to chassis setup, vehicle dynamics and race car design, together with solid engineering theory.

This is not a controversial race car setup book and agrees in principle with technology and theory taught in major college motorsports programs. To many, according to what I hear, the books of mine that have preceded this one have become their bible of racing knowledge. Much of the technology presented here has been more recently developed over just the last five years or so. For those who believe that we had already pushed the envelope of vehicle dynamics about as far as it could go by the early 2,000's, this book does not follow that line of thinking.

Regardless of what is on the pages, the proof is on the racetrack, and the methods in this book have been tested and proven to improve performance in race cars. They increase speed, improve basic stability and have already been used to win many races and championships in many classes of racing.

How then did these RCT books come about? When I began my career working with race cars, I found plenty of information on chassis theory, but I couldn't find conclusive information that would tell me how to set up my race car in the shop the right way the first time. I had to read between the lines and keep trying different setups, working by trial and error. I personally don't like trial and error. I want to be able to know exactly how to set up my race car and know not only how something works, but why.

This book will help you avoid the trial-and-error approach to chassis setup. It will teach you sound, proven technology that is both easy to understand and easy to use, so you can set up your race car in the shop and see the positive results on the track immediately, with very little tweaking. What follows is a common-sense approach to chassis setup, vehicle dynamics and race-car design, founded on solid engineering theory. However, you will need to have an open mind, and be willing to accept new ideas that may go against previous chassis setup thinking.

Just to make it clear, the technology presented here applies to all race cars, from quarter midgets to Formula One and everything in between. This book tends to lean towards stock car racing because it represents most of the world's automobile racing. But know that not only will be useful for all forms of circle track racing from asphalt types to dirt cars, a great deal of the technology applies to all race cars.

Success comes at all levels of endeavor, and we can't all be champions. But we can all get better at what we do. The goal of this book is to give good, solid information that has been tested and evaluated and found to be the truth. It is not, and will never be, complete as long as we continue to push the envelope in the search for better performance, but it will lay the foundation upon which future race engineers can build their programs.

Race Car Technology – Level One

Welcome To Race Car Technology - Level One

Thank you for enrolling in the RCT Level One course. You have taken the first step in furthering your knowledge and understanding of the systems and parts of a dedicated race car. This course will explain all of the relevant systems race teams and race engineers work with on a daily basis. This course will teach you how to speak the language of race car technology.

Level One provides the foundation for the next three Levels in the Race Car Technology school. There is information contained in this Level that has never been presented before in any format. For that very reason, even seasoned veteran racers will come to appreciate the presentation of this unique knowledge.

The information presented is in some ways about specific race car types, and then many ways, much of what you will learn is relevant to all race car types. The basic technology is common among race cars from simple street stock circle track cars to sophisticated road racing and formula cars.

A very sophisticated formula car shares many common traits with less sophisticated race cars. The part and pieces are constructed differently and the costs are much different, but all in all, we are talking about a race car. With this Course you will be able to piece together a race car part by part so that you will come to fully understand how they all work.

The understanding of, and the setup for, any race car depends on certain common knowledge of all of the parts of the race car, and how all of those parts interact. Whether it involves a late model circle track car, or a world class formula car, what you will learn here will be relevant to any and all race cars.

Please know that you can proceed at your own pace. This is not a race and you will get no better grade for finishing quickly. In fact, it is much better to take your time and try to fully understand the concepts and information contained in each of the Lessons presented here.

The Lessons are arranged in a predetermined order so that you can piece this all together at the end. In fact, the order in some cases is critical so that the information builds on what has previously been discussed.

There will be a test at the end of each Lesson and a Final Exam at the end of the course. These exams are necessary to ensure that you have grasped and learned all of the information presented here that will allow you to earn your Certificate of Completion upon completion of the three Levels for Race Car Technology. By passing these exams, you will have proven to yourself and to others that you have retained important knowledge about race car technology.

So, let's get started. As you go to Lesson One, again remember to take your time and repeat reading each section as many times as it takes to understand the concepts and information presented. Good luck and let's get started.

Race Car Technology – Level One
Lesson One – Goals: What We Trying To Achieve

The Race Car Technology – Level One Course is designed to introduce the student to the parts and pieces of a race car. We will explain what the systems are, what they are used for, and how these systems work with the other components on the race car. Many of these systems are common among various race car types and uses. Once you have completed Level One, you will know what the systems are, and what to call the parts used to make up the systems. This is an important and necessary course leading into our RCT Level Two Course, which takes us to the actual setting up of the race car.

Before we get into the nuts and bolts of race car technology, we need to fully understand what our goals are for this school and exactly what we are trying to achieve in RCT Level One. For every racer, no matter what class, type, sanction or level of expertise, the goal is simply to win races. To win races you need to be faster than any of the other competitors. To be faster, you need knowledge. We are going to give you some important knowledge.

The beginning of knowledge, someone once said, is the admitting that you don't know. Not knowing says you do know there is knowledge out there that you have not yet been introduced to. Being open to this philosophy helps you to seek and gain knowledge throughout your journey in life.

For any student to proceed through a course, you need to be aware of and understand the words that are going to be used in the process, and the meaning of those terms. You can always refer to the Glossary if you forget or need to remind yourself.

The race engineer's goals are very simple in concept, but more complicated to carry out. On any race course, be it with circle track or road racing, on a dirt or asphalt surface, there are key areas of performance where your car needs to be made better than any of your competition and those are the following;

The goals for designing our race cars are basically the same be it a formula road racing car or a late model circle track car. Performance gains in the slower portions of the track cause all of the speeds to increase around the track including on the straights.

Maximizing Grip - When non-chassis elements like powerplants are more or less equal, Grip is what will help you win races and championships. Motors accelerate you, brakes slow you, but Grip makes you faster through the slower turn portions of the track and relatively small gains in lateral Grip will produce huge gains in speed and performance.

What are the Key elements of Grip? The tire contact patch is where Grip is produced. Causing the tire contact patch to be larger increases Grip and causing the most vertical loading on that tire contact patch are the two key ingredients for maximizing Grip.

Maximizing the amount of traction that is available from the four tires on a race car, any race car, will make you as fast as you can be, all other things being equal. Everything we present in RCT L1 will ultimately lead to optimization of the race cars Grip and the use of that Grip to go faster.

What are the basic parts that make up Grip?

• Loading On The Tire – The more load we can put on a tire, the more Grip that tire will have, period. But, this is gain in grip is not linear as we will explain later on. Load can come from the weight of the car, mechanical downforce from banking, and aerodynamic downforce, all of which will be explained in detail later on.

• Contact Patch Area – The greater the size of the contact patch, the more Grip. If we can find ways to make the tire contact patch larger, then that tire will produce more Grip.

• Tire Compound – The physical and chemical makeup of the tire can provide more Grip. The softer the material, the more Grip within limits. We have rules that govern the softness of the tires, but we need to stay very close to those limits.

• Load Distribution – A pair of tires on the same end of the car, or same axle, will produce more Grip when they are more equally loaded. The most Grip from a pair of opposing tires will come when they are equally loaded. There is a variation to this concept for dirt racing that will be addressed later on.

• Angle of Attack – What is called Angle of Attack, or Slip Angle, is when a tire is pointed slightly to the inside of the arc it is following through a turn. If a tire were to follow the exact tangent line around a curve or arc of the turn, it would not generate any side force to counter the centrifugal force.

So, in consideration of the other items that make up Grip, it is fair to say that none of those would be useful if it weren't for the creation of Angle of Attack. No matter what amount of Loading or size of the Contact Patch area is, or how soft the tire compound is, or how equal the load distribution between opposing tires, the car would not stay on the course without the tires developing an Angle of Attack.

So, there you have it. Those five things represent the parts that help make the Grip we seek to make us faster so we can win races. As we go through each part of the race car in the other Lessons, we will explain how to optimize of those parts to enhance our Grip and make us faster through the slow speed turns. And we will understand how that will in-turn make our whole lap faster at every point around the course.

Why Does More Grip Make Us Faster?

When a race car turns, a lateral force called Centrifugal Force tries to push the car to the outside of the turn. The tire contact patches resist this force. The speed we can drive through the turn is limited by the amount of Grip we have in our tires. The more Grip, the faster we can drive through the turns.

The one often overlooked benefit of achieving faster turn speeds is this: the faster you exit a slow speed turn, the higher the speed at which you will start accelerating down the straight part of the track in-between the turns. So, speed gained in the slow speed turns will be carried down the straights too. It's not just that we gain speed in the slow speed turns with more Grip, we gain everywhere around a race track.

To give you an example, on a typical or average length Formula One track, one tenth of a second is about 17 feet on the race track. Some teams are a full second slower per lap than the fastest teams. That is 170 feet per lap that the faster cars move ahead of the slower cars each lap. In a fifty lap race, that equals 1.6 miles, or 2.5 kilometers.

We often think of the formula one cars as being the ultimate race cars and perfect in every way. If that were so, then there wouldn't be a two second gap between the fastest and slowest cars. There must be a difference in designs that causes that huge gap in speeds. We like to think it is lack of mechanical grip that makes not only the slow turns faster, but the entire lap.

We already know that some teams get lapped in a F1 race, and the average length of those tracks is around 3 miles. So, those lapped cars are on average 2.0 seconds per lap slower than the winning team. And they all run with the same choice of tires. That is hard to fathom.

Is the gain realized by the winning F1 teams all Grip? No, as we have read, some engine packages are down on horsepower and that is a factor in the difference in lap times. But what about the teams who have the same engine package as the winners? Why are they so slow? It could be that they lack mechanical grip in the slower portions of the race track.

Building mechanical grip is one of the most important areas of race car engineering, even over and above aerodynamic grip. The reason is this; To gain maximum aero grip, you must have high speeds and the most gains from mechanical grip happen in the slower turns where the speeds and the aero downforce is lower for all of the teams. So, the only thing we can point to as the reason for the gain in performance for the winning teams is Mechanical Grip.

The Concept Of Mechanical Balance - If your car has more overall Grip than other cars early in the race, that Grip might not stay superior and you might end up being a slower car through the turns later in the race. This is due to the Balance Factor. There are two definitions of balance. They are not the same and they cannot be confused. One is handling balance which is defined as a car that is neither pushing (understeering) or loose (oversteering). Pushing is when the front set of tires has less Grip than the rear set of tires. Loose is the opposite.

Circle track race cars are complicated, but much easier to setup that a road racing car. This is because the setup in a circle track car can be asymmetrical, meaning the spring rates, cambers, etc. from side to side can be different. The circle track car is only turning in one direction and so the setup need only be correct for that direction of turning.

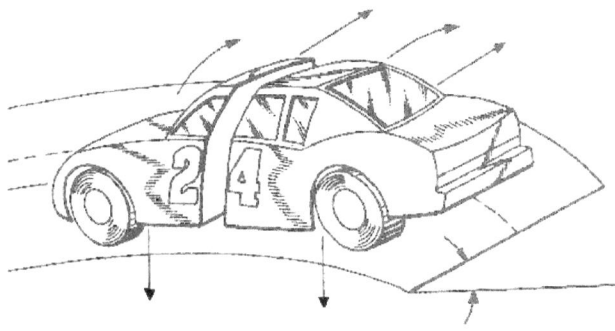

The Mechanical Balance concept is simple. We want both ends of the car to work in sync so that one does not try to cause the other to do what it does not want to do. When the two ends of the car are working in unison, we can achieve the goal of mechanical balance.

A team can attain handling balance fairly easily, but just the fact that the car is neutral in handling doesn't translate to speed or consistency. A car can have less overall Grip than other cars and still be neutral in handling. Although a neutral handling car is a goal, it is not the first and primary goal and not what we are discussing here.

The other balance is Mechanical Balance. Simply put, this is when the two suspension systems, front and rear, are working in sync and where the load transfer is predictable and maximized. This is what teams need to search for. The end result of all we do with chassis engineering must be towards the goal of Mechanical Balance.

The Search and Desire for Mechanical Balance should be the foremost goal for every race engineer. A lot of good things come with achieving MB. The tires do not work as hard and will last longer. The car is much easier to drive. The car maintains higher turn speeds as the tires wear later in the race, or later in each stint for cars that are allowed to pit and take on new tires.

So, to reiterate what we have covered so far, our goals for this course are to gain knowledge, understand the terminology, and to learn how to setup a race car to maximize Grip in a way that utilizes a Mechanical Balanced state. In the Lessons to come, we will be presenting information you will need to understand and apply principles that will help achieve these two important goals.

And we are not leaving aerodynamic downforce out of the discussion, we just need to separate the mechanical Grip from Aerodynamic enhanced Grip. We need to do this because, many times in the search for the maximum aero properties, a team might deviate from the goal to achieve a Mechanical Balance. We think you can have the best of both worlds.

Exam In The Context Of This Lesson:

What is the beginning of knowledge?
1) Reading books
2) Working on a race team
3) Asking Questions
4) Admitting You Don't Know Everything

What wins you races and championships?
1) Superior Grip
2) Handling Balance
3) Mechanical Balance
4) All of the above

Which of these are basic Elements of Grip?
1) Tire Pressures
2) Tire Temperatures
3) Contact Patch Area
4) Loading
5) 3 and 4

What is the name of the force that tries to push the car off the turns?
1) Downforce
2) Drag
3) Centripetal
4) Centrifugal
5) Gravity

Low speed corner gains from Mechanical Grip increases performance where?
1) Down the straights
2) In the high-speed turns
3) At the start of the race
4) At the end of the race
5) All of the above

The first and primary goal for performance in a race car is?
1) Equal weights on all four tires
2) A car that is neither tight nor loose
3) Mechanical Balance
4) Aerodynamic Downforce

Lesson Two - Why Do Race Cars Turn?

- ***Tangent Line of An Arc*** – The tangent of an arc is a line that is at right angles to a line from a point on the arc to the radius. This tangent forms a "T".

- ***Angle of Attack*** – A direction of a tire that is pointed towards the inside of the turn from the tangent line. This is commonly referred to as a Slip Angle in most technical textbooks.

What we are about to explain is a concept that has never been fully explained as far as we know. It is a simple, but effective, description of why a car can drive through a turn at high speeds. It is one of those items of new knowledge that we promised the veteran racers they would get when considering enrolling in RCT Courses.

In Lesson One, we talked about how a tire develops Grip. One of the key ways is from Angle of Attack or Slip Angle. A tire running parallel to the path of the arc of the curve cannot generate the Grip it needs to maintain its direction along the arc. It would "slip" off the arc.

Every race car in the world, or 99.999 percent of them, is a front-end steering car. That is, the driver turns the car using a steering mechanism that is used to change the direction and angle of the front wheels.

In this way, the driver creates an angle of attack for the front tires so they can develop the side force necessary to counter the centrifugal force. Without this Angle of Attack, the car would not turn no matter how much tire loading, or size of the tire contact patch there was.

But what about the rear tires? We cannot steer them, setting aside the effects of mechanical rear steer that will be covered later on. How do they develop the angle of attack needed to stay on the arc of the turn? The answer is: they already have their angle of attack, and here's why.

In Figure 2-1 you will see how each tire follows its own arc. The arc radius of the inside (inside of the turn) tires is less than that of the outside (outside of the turn) tires. In Figure 2-2, each tire is placed over an exaggeration of the arc lines with a radius equal to the distance from each set of tires to the common radius point of the arcs.

In the Figure 2-2 sketch, all four tires are pointed straight ahead to show their relationship to the direction of the tangent line. One arc is for the inside, or left side tires and one is for the outside, or right side tires in this example.

As you can clearly see, at the front, the tires are pointed away from the tangent line. For those tires to turn the car, not only do they need to turn left to the direction of the tangent line, they need to go a little bit further to create the angle of attack needed to develop the amount of Grip needed to stay on course.

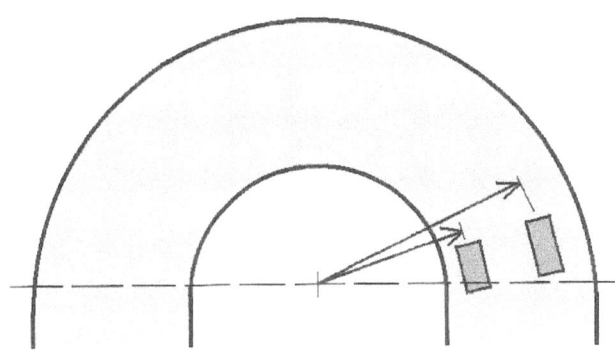

Figure 2-1

In this illustration, we show how the inside and outside tires have different radii and must follow different arcs when driving through the turn. A line perpendicular to the line extending to the radius point is called a Tangent. To develop the Grip needed to counteract Centrifugal force, all of the tires must be pointed to the left of the Tangent line.

For wheels that can be steered, this is easy. The driver just need to turn the steering wheel in that direction. And the driver will do that to the extent and to the angle needed to turn the car and follow the line they think they need to run through the turns. Sounds pretty simple, right?

Now let's look at the back tires. The direction they are pointed is to the left of the tangent line. Unlike the front tires, they already have an angle of attack, whether or not what it has is exactly what is needed to stay on track. Interesting, right? It gets better.

Now let's say there's not enough angle of attack in the rear tires to keep the tires on course and the rear tires slip a bit. Guess what, as the rear tires slip, the angle of attack increases. In reality, for a race car, the rear will slip if it doesn't have quite the Grip it needs until the needed angle of attack is reached and then it will stop slipping. It is therefore self-compensating.

The front tires react the opposite if they slip. The angle of attack decreases and the steering must be increased in order for the car to stay on the course. We will tell you more about steering angles and angles of attack later on. There are limits to what angle the front tires can be steered to.

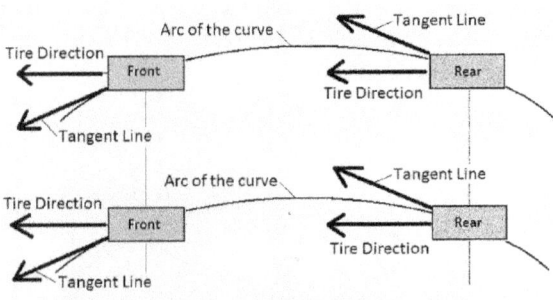

Figure 2-2

The front and rear tires are pointed in different directions in relation to the Tangent of the arc they are following. In the front, the tires are pointed away from the Tangent and in the wrong direction needed to turn the car.

We must steer those tires left in this left turn in order to not only track parallel to the Tangent line, but a bit more to develop an Angle of Attack needed to generate the amount of Grip necessary to keep the car on the course.

At the rear, we have a different situation. With the rear tires positioned straight ahead, they are already pointed left of the Tangent line and have some Angle of Attack.

If the angle is less than needed, the rear of the car will slip out until the tire angle is enough to provide sufficient Angle of Attack needed to stay on the course.

Here we can actually see the lines of the arc of the turns. The front wheels are steered to create the Angle of Attack needed to develop the force to counter the G-lateral forces. The rear tires are already pointed to the inside of the turn and have their own Angle of Attack. This is why race cars turn the corner.

Know that every race car, no matter the design or classification, is subject to the concept explained above. Farther along in this Course, we will get into various design effects, goals and geometry for the rear suspension whereby we can change and fine tune the angle of attack for the rear tires.

Although this Lesson is relatively short, it is vital to understanding much of the coming presentations about vehicle setup and dynamics and the parts and pieces of the race car that make all of that happen. We'll keep referring back to this concept in Lessons to come.

When you come to understand this concept of how the car is able to makes it around the turns, it will answer a lot of questions people might ask about vehicle dynamics. I have often wondered, given that we have a lot of setup tools to make the front ends turn, why and how does the rear turn with the front? Now we know.

Exam - In The Context Of This Lesson:

Which Element of Grip Is Essential Above All Others?

1) Tire Loading

2) Tire Compound

3) Angle of Attack

4) Contact Patch Area

How Do We Make The Front Tires Develop Angle of Attack?

1) Add More Camber

2) Add Tire Pressure

3) Drive Slower

4) Turn the Steering Wheel

How Do We Make The Rear Tires Develop Angle of Attack?

1) Add More Camber

2) Add Tire Pressure

3) Drive Slower

4) We don't need to, they already have Angle of Attack

How Do We Adjust The Angle of Attack For The Front Wheels?

1) Lengthen the Tie Rods

2) Change The Speed

3) Change The Tire Cambers

4) Turn the Steering Wheel Either Way

How Do We Adjust The Angle of Attack For The Rear Wheels?

1) Change The Tire Stagger

2) Change The Tire Pressure

3) Adjust The Speed

4) We Don't, The Rear Is Self-adjusting

Lesson Three – Cambers: Definition and Effect
The Front Geometry Group

When we start to develop our Grip, we start with the tires. There are several components related to tire optimization, or making the tire happy. Camber is one of those components.

Camber Is - Simply put, Camber is the tilting of the tire/wheel combination either with the top of the tire in towards the center line of the chassis, or out away from the centerline of the chassis. It is measured in degrees of tilt from vertical relative to the racing surface.

When the top of the tire is tilted in, that is called Negative Camber. When the top is tilted out, that is called Positive Camber. Wheel Cambers are adjustable in most race cars. And wheel cambers can change with chassis motion.

For this discussion, we will be talking only about cambers in a double A-arm suspension. Stock cars mostly have only front AA-arm suspensions and a solid axle rear suspension. Many prototype racing cars and formula cars have independent AA-arm front and rear suspensions.

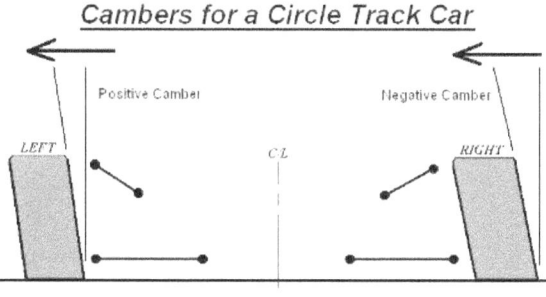

The reason why we want and need camber is because the tire distorts as lateral loading is applied to it when cornering. This distortion changes the contact patch shape and may cause the tire to have less Grip. The outside tire (meaning outside of the turns away from the radius) carries more load in a turn and therefore needs more camber than the inside tire.

So, we add negative camber (top tilts in) to the outside front wheel. For a circle track car turning in only one direction, we add negative camber to the outside wheel and positive camber to the inside wheel.

For road racing cars, we need to set negative camber to both front wheels since we will be turning left and right and in a right hand turn the left front tire will need negative camber just as in a left hand turn the right front tire will need negative camber.

The amount of camber that a particular tire needs is dependent on three factors. These are: 1) The tire composition, mainly the side wall stiffness (softer sidewalls require more camber), 2) the amount of lateral loading (the more loading the more the tire will roll over and then the more camber we will need), and 3) the camber change characteristics of the suspension.

So, let's recap. A tire needs camber to correct what the tire contact patch does when cornering with a high lateral load. Since the tire distorts, the camber serves to correct the situation and provide a more ideal contact patch shape.

A tire will distort to the inside of the turn since the rest of the tire is trying to move out away from the turn. In circle track racing, we apply positive camber to the inside tire and negative camber to the outside tire.

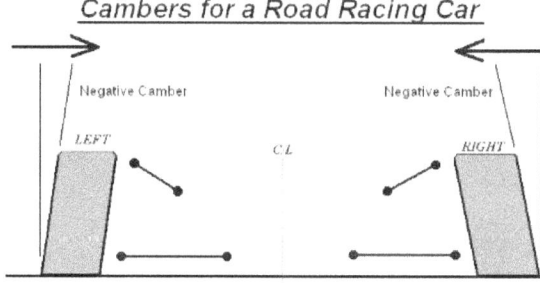

In road racing, both tires will be the "outside" tire depending on which way we are turning, so we cannot put positive camber in either wheel.

Let's say we took a circle track car, that only turns left, to a road course. It would turn very well in the left turns, but in the right turns, both of the tires would have the opposite of the direction of camber it needs to correct the tire distortion.

Again, in road racing we sacrifice the tire contact patch of the inside tire to benefit the outside tire that will be carrying the most loading through the turns. And there may be ways to minimize that sacrifice that will be explained later.

Camber Change – In a modern AA-arm suspension, as the chassis dives and rolls through the turn, because of the design of most AA-arm suspensions, the wheel camber will change relative to the racing surface.

Camber Change can be designed differently to optimize or minimize the camber change for a particular chassis design.

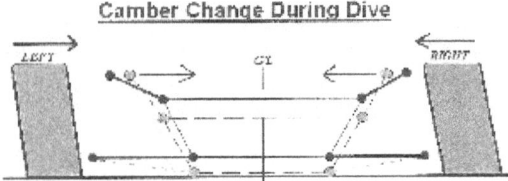

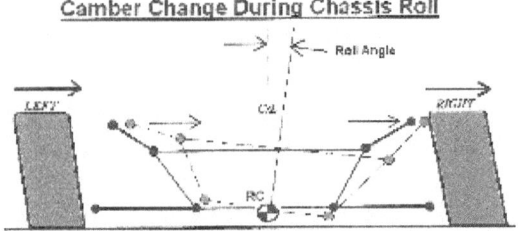

Camber Change is related to the combination of motions of the chassis. In a turn, the chassis moves. Most of the time, in all race cars, it gets lower and it rolls.

It has been common in the past to talk about camber change only related to vertical chassis movement. A team might set the car at ride height on supports and then move the wheel up to see how much the camber changes in degrees per inch of movement. This is camber change, but only from one single part of the movement of the chassis.

We can measure the static camber in our wheels by using a bubble level gauge and directly read the camber amount.

We can also measure the static camber in our wheels by using a digital gauge that reads the camber amount in degrees and decimals.

For the purpose of this Course, and to look at camber change in a more realistic way, we will only be thinking about camber change when it is related to both the vertical chassis movement which is combined with chassis roll. Here is why.

If we roll the chassis without moving it vertically (i.e. the center remains at the same height), the front wheel cambers will change the same amount in degrees as the chassis rolls. If we move the wheels vertically without rolling the chassis, the wheel cambers will change some amount, but in the opposite direction. So, if we combine the movements, we can then determine what the net camber change is relative to the racing surface. That is what matters.

In extreme cases, with very soft sidewalls, it might be an advantage to run very high amounts of negative camber in the outside wheel in order to generate the greatest tire contact patch amount.

Designing Camber Change – We can change the amount of camber change in some race cars. If you can make changes to the arm lengths and/or arm angles, then you can change the amount of camber change.

Installing shorter or longer upper control arms, or making changes to the angles of the upper control arms will change the amount of camber change.

In most cases, race tires do not like camber change. When entering a turn, the tire takes a set, or rolls over and begins to develop Grip. If the cambers do not remain stable, the tire will take more time to adjust, and frankly, we don't have a lot of time for that.

The result is loss of Grip and loss speed as the tires struggles to find Grip. So, in modern race car design, we need to plan out our AA-arm geometry so that there will be minimum camber change. How do we do that? There are several ways to accomplish the goal of minimum camber change.

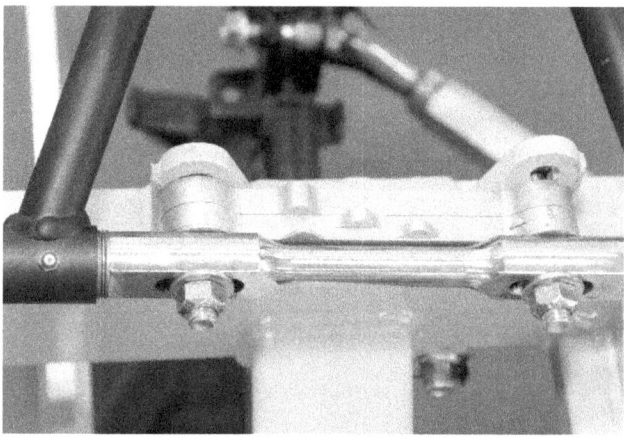

Camber is easily adjusted for this circle track car by changing the spacers between the frame mounting plate and the upper control arm shaft. For formula cars, the cambers are changed in a similar way, by moving the inner mounts of the upper control arms or extending the length of the upper control arms.

First, we can arrange the control arms with angles so that with the dive and roll we expect going into and through the turns, will result in minimum camber change. This has been done with great success in what we call conventional setups with both stock cars and some formula cars.

Another way to achieve minimum camber change is to reduce the amount of dive and roll. The setups in Formula One cars are very stiff and the chassis does not move vertically very much. This reduces the amount of camber change. The reverse angle in an F1 car (meaning that the inside chassis mount is higher than the ball joint) of the upper control arms may be beneficial and cause a small amount of beneficial camber change. We will get more into that in the Level Two RCT course.

The third way we can minimize camber change is through the use of what we call bump stops. In modern race car design, some chassis are designed to travel a predetermined amount and then the shocks contact a bump material and basically stop motion.

These bumps can be made of hard rubber, polypropylene or actual steel springs. In each case, when the shock contacts the bump device, the vertical chassis motion is reduced to very small movements and the cambers change very little while the suspension is on the bumps.

Cars that use bumps include dirt late models, asphalt late models, stock car touring divisions, IMSA prototypes and Formula One. There are other benefits to using the bump setups, but reducing camber change is one of the most significant benefits realized by running on bumps.

Ultimate Camber – The Static camber settings (without dive and roll with the car is at ride height and not moving) are something teams experiment with to try to get the best Dynamic (after dive and roll) cambers that will provide the most Grip. Because all car designs and setups are different, so too are the camber changes for each car.

The ultimate Static cambers for one race car might be different than what is optimum for another in the same class using the same tires. The differences are caused by: 1) Variations in tire pressures, 2) Different front geometry designs that result in different camber change from similar chassis movements, and 3) Different stiffness in setups that will cause different dive and roll amounts.

We are going to cover suspension design, tire technology, and chassis setup dealing with stiffness in later Lessons in RCT Level One, but we want you to understand the basic characteristics of camber and camber change going forward.

Exam - In The Context Of This Lesson:

Tire/Wheel Camber Is The Degree of Tilting Of The Wheel From What?

1) Horizontal

2) A line at right angles to the racing surface

3) A Vertical Plumb Bob

4) A line through the ball joints

Negative Camber Is?

1) When It Moves The Wrong Direction

2) The Top Is Father Out Than The Bottom

3) When It Doesn't Change At All

4) The Top Is Closer To The Centerline of the Car

We Need Camber Because?

1) The Tire Distorts When Lateral Force Is Applied

2) The Suspension Points Move In Dive And Roll

3) Camber Helps Generate More Lateral Force

4) All Of The Above

Camber Change Occurs Because?

1) The Wheel Moves Up and Down

2) The Car Rolls

3) The Tire Distorts From Lateral Force

4) A Combination of 1 and 2.

What Changes To The Suspension Will Affect The Amount of Camber Change?

1) Changing the Spring Rates

2) Installing different length control arms

3) Changing the angles of the control arms

4) All of the above

How Much Camber Changes Do We Need?

1) Enough to get what the tire needs

2) An amount equal to the tire roll over

3) As much as possible

4) As little as possible

What Are Common Ways To Minimize Camber Change?

1) By running very stiff setups

2) Through special design of control arm angles

3) With the use of bump devices that limit suspension movement

4) All of the above

The Ultimate Cambers We Need Are Determined By?

1) The Amount of Lateral Load

2) The Degree of Camber Change In Our Suspension

3) The Stiffness of the Tire Sidewall

4) All Of The Above

Lesson Four – Caster: Definition and Effect

Caster and Camber are often talked about together, but they are much different in their design and use. And Caster can cause camber change, but that is not its normal use.

Caster is a design function in a vehicle to promote directional stability. It helps the vehicle to want to roll straight ahead and not deviate from a straight line.

Caster is used in many different types of vehicles, even grocery carts. In bicycles and motorcycles, it is the slant of the forks where the front wheel axle is well forward of the fork tube just below the handlebars. When weight is put on the bike, the wheel wants to point straight ahead.

If we turn the bike wheel, it tends to want to raise the bike. This goes against or opposite of the weight pushing down on the wheel from the weight of the bike plus your body weight. The weight and angle of the forks will resist any wheel movement from straight ahead. That is what caster does in a bike, and that is what caster does in a race cars front steering system.

What Caster Looks Like - Caster is set in a double A-arm front suspension by positioning the ball joints at an angle looking at them from a side view. There is Positive and Negative Caster.

In the side view, if the top ball joint is to the rear of the car in relation to the bottom ball joint, we call this Positive Caster. If the upper ball joint is forward in relation to the lower ball joint, we then have Negative Caster.

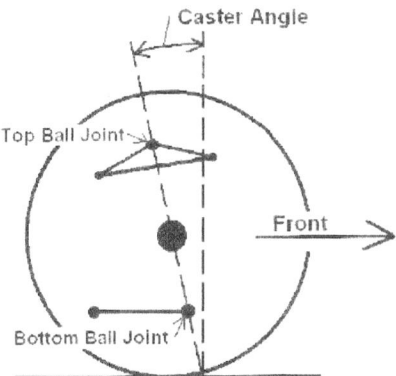

Caster is created when we design the suspension in a way that places the upper ball joint to the rear of the lower ball joint. This is called Positive Caster. If the upper ball joint were to the front of the lower ball joint, then we would call that Negative Caster.

Caster is measured in degrees of caster. This is the exact number of degrees that a line through the upper and lower ball joints centers of rotation makes with a vertical line as viewed from the side. It can be measured directly or with a tool called a camber gauge. We'll explain how to use this gauge later on.

Caster in a race car can make turning the steering wheel harder. The more caster, the harder it is to turn the steering wheel. With the use of power steering in a race car, this effect is reduced or eliminated.

How Much Caster? - We usually set the caster amounts, or degree of caster, the same for each front wheel in a road racing situation. The trend for road racing cars is to set high amounts of caster and it is thought that this enhances other effects of caster. More on that later.

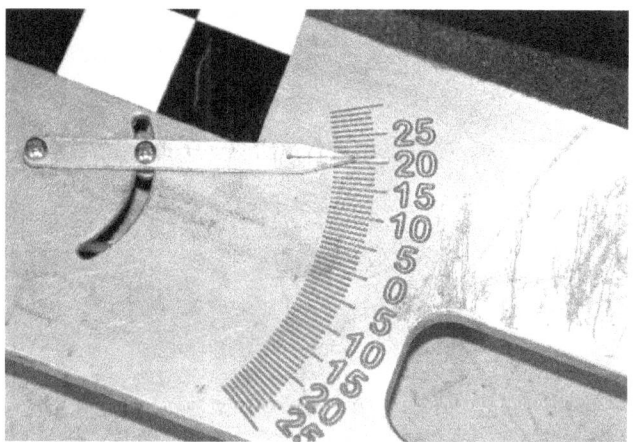

To measure caster, we first need to use a turn plate or other means to establish the number of degrees we are turning the wheel. The accepted SAE, or Society of Automotive Engineers, method is to turn the wheel twenty degrees one way from straight ahead and note the camber reading. Then we turn the wheel to twenty degrees in the opposite direction from straight ahead and remeasure the camber. The difference in camber amounts is the caster.

For circle track racing where the cars always turn one way, the caster settings for the inside wheel is made less than what we set for the outside wheel. The inside wheel caster may even be Negative, or with the upper ball joint forward of the lower ball joint. This is commonly called Caster Split or even Caster Stagger.

We stagger the caster settings in circle track racing to cause the car to want to turn into the corner. For circle track racing in the U.S. where the car turn left, the caster is always less on the left side than the right side, if you were to release the steering wheel when rolling forward, the car would naturally turn left on its own. The more caster split we design into the suspension, the more it will want to turn left.

Caster Changes - Caster Change is caused by wheel movement when there is a difference in upper and lower control arm angles when viewed from the side. If the upper control arm is angled from the side view and the lower control arm is either level or angled the opposite way, the caster will change when the spindle moves vertically.

If you use a camber gauge, when you turn the wheel out twenty degrees, you set zero in the gauge as shown. This is the right front wheel on a stock car and it is turned out.

We then turn the wheel in twenty degrees from straight ahead and re-read the camber gauge and this tells us the amount of caster.

When there is a difference in side view angles of the control arms associated with one side of a AA-arm suspension, the ball joints will move forward or rearward relative to each other and that changes the side view angle of the line running through the ball joint centers which defines caster.

When this line changes its angle, then the caster is also changing. We need to be aware of these changes when designing our race cars. If the right caster is increasing with vertical movement and the left caster is decreasing, or even staying the same, we are gaining caster split and the increasing the effect of the caster split.

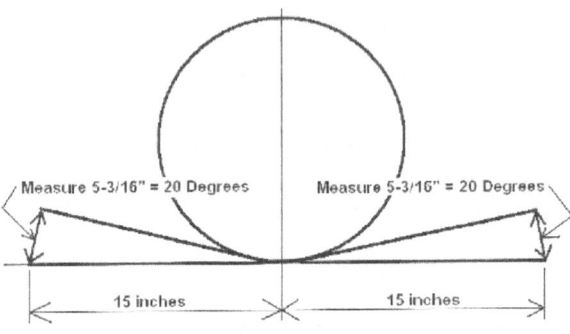

When you measure caster using a gauge, you need to know how far 20 degrees of rotation is. If you don't have turn plates to read the degrees of rotation, then this sketch shows you how to mark the 20 degree line on the floor so that your turning the tire the correct number of degrees.

Camber Change Due To Caster - As we turn the steering wheel, caster causes a change in camber for each wheel. The amount of camber change depends on the amount of steering and the amount of caster.

If we turned the wheel both ways 90 degrees from straight ahead, the camber would change the exact number of degrees as the caster amount. This is how caster gauges measure caster. They are designed to record in a similar way to what we described above. With the gauge, we turn the wheel out from straight ahead by exactly twenty degrees, set zero in the gauge, then turn the wheel to be pointed in from straight ahead and then read the gauge. It will read the degree of caster.

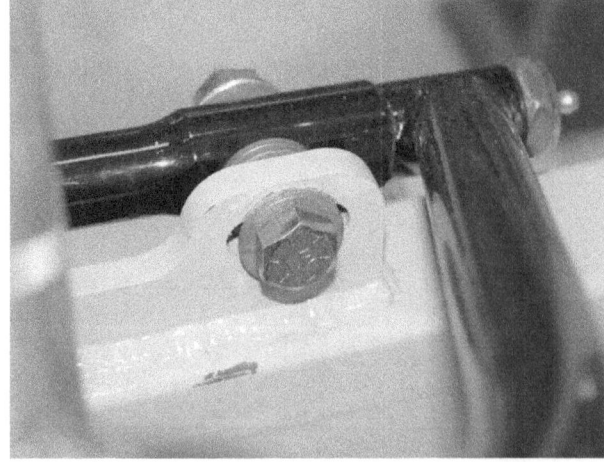

This upper control arm mount has a slot in the mounting plate that allows changes to the caster. By sliding the upper control arm shaft fore or aft, you can reduce or gain caster.

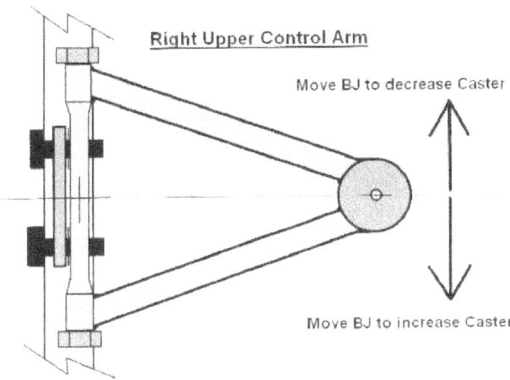

We make changes to caster in a stock car by moving the upper ball joint forward or rearward. Moving it to the front reduces the caster and moving it towards the back of the car increases the caster. Teams can install upper control arms with offset ball joint mounts to create the desired amount of caster.

There are other effects associated with caster, but we will cover those in our Lesson on Spindles later on. But to re-cap for you, Caster is a self-aligning effect that helps keep the wheels pointed straight ahead. Caster is caused by mounting the ball joints so that the upper ball joint is positioned to the rear of the bottom ball joint. This is referred to as Positive Caster.

If we have different stagger amounts on each side of the car, the car will want to steer in the direction of least caster. Running less caster in the left front of a left turning circle track car will cause the car to want to steer left on its own, without driver input.

And, caster causes camber change when the wheel is turned, along with other settings in the front suspension. The combined net amount of camber change is dependent on the other factors in the design of the suspension.

Exam - In The Context Of This Lesson:

Caster Is The Degree of Tilting Of The Ball Joints From What?

1) Horizontal In A Front View
2) Horizontal In A Side View
3) Equal to Camber
4) Vertical In A Side View

We Need Caster Because?

1) It makes the car turn easier
2) It adds directional stability
3) It helps the tires gain Grip
4) 1 and 2

Caster Is Measured By Doing This?

1) Measuring the height of the ball joints
2) Knowing the control arm angles
3) Measuring the degrees of sideview angle formed by the ball joints from vertical
4) Using a caster gauge
5) 3 and 4

Caster Split Is Caused By?

1) The change in caster from wheel movement
2) The movement of the spindles
3) The difference in the left and right caster amounts
4) Turning the steering wheel

Caster Change Is Caused By?

1) The lengths of the control arms
2) Turning the steering wheel
3) Wheel movement with differences in sideview control arm angles
4) Chassis dive and roll

Lesson Five – Spindle Design

The Spindle is the connection between the chassis and the wheel acting through the control arms. The spindle is a solid component that attaches to the upper and lower control arms through two ball joints, upper and lower, that are free to rotate in three dimensions to some extent.

The wheel is attached to the spindle by being mounted on the spindle pin. On this pin are mounted two sets of bearings that are connected to bearing races within the wheel hub. This completes the connection between the chassis and tire contact patch.

The spindle can be designed in many different ways to affect camber change, caster and caster trail, and primarily scrub radius.

The modern fabricated spindle used on circle track and road racing cars is made from welding flat plate steel into the shape needed for the application. These pieces are lighter in weight than the older cast iron spindles.

This is a common, stock type of spindle (right of the shock and left of the orange colored brake rotor) that attaches to an upper and lower ball joint. The other connection is the tie rod end that is bolted to the spindle steering arm shown at the very front of the spindle.

Spindles can be made out of any material that is practical, like this one made from carbon fiber using aluminum bolt on parts to hold the wheel spindle pin, ball joints and steering links.

19

The variations in spindle design are made to accommodate different race car designs. There are three basic design aspects for a spindle.

- Ball Joint Spacing – the distance between the ball joints for each spindle design is set by the spacing between the ball joint mounting holes is a stock car spindle, or the spacing between the heim joint mounts in a formula type of race car.

- Spindle Inclination – When the spindle is viewed from the front, the ball joints or centers of heim joints are staggered from a vertical line. In every case, the upper ball joint is placed towards the center of the chassis more so than the lower ball joint. The line through the ball joints when viewed from the front is the Spindle Inclination or what is sometimes called king pin angle, a reference back to the days of straight axle cars. This angle can vary depending on the design goals for each race car.

- Spindle Pin Height – The height of the spindle pin in relation to the positions of the ball joints varies depending on the design goals for each race car.

The spindle on this prototype is made from high strength aluminum. The upper control arm is mounted to the spindle by way of a mono-ball joint, in place of the larger, stock type ball joint. We can also see the steering arm forward of the upper mono-ball that attaches to the steering tie rod.

Scrub Radius Defined - Scrub Radius occurs when a spindle is designed whereby the point where a line through the centers of rotation of the ball joints intersects with the driving surface is some distance from the center of the contact patch. When a force is applied to the tire such as braking, if there is a scrub radius, then that force will want to turn the spindle.

The design goal of ending up with minimum Scrub Radius comes from early designs of production automobiles. A wheel assembly (spindle, wheel offset, wheel diameter, etc.) that is designed with a large amount of scrub radius will produce a moment arm of sorts. If the vehicle were to hit a pot hole, a force would be created which tries to rotate the wheel backwards.

If the wheel assembly had some amount of scrub radius, then the wheel would in fact be pushed back and this would feed a force through the steering system and the driver would feel that. This was not a desirable condition and could cause the car to steer violently in the direction the wheels were forced to move to.

By designing a wheel assembly that had zero scrub radius, this undesirable condition could be eliminated. Now, in modern day racing, scrub radius is not considered an important design criteria. Modern race tracks don't have pot holes.

All of this sounds complicated, but it's not. Just remember that scrub radius is not a performance enhancing component of a modern race car. With the advent of smooth race tracks, even on dirt, and power steering systems, even if we do hit bumps, the feedback to the driver is minimal is any at all.

Ball Joint Location and Spacing - So, we need to concentrate on those parts of the spindle that do relate to performance. The car is supported, in part, by the spindle. Most race cars have the spring attached to the lower control arm and that extends out to the spindle. So, end of the lower control arm attaches to a ball joint that is mounted to the spindle.

Since the spindle pin is basically the center of the wheel, its position in relation to the lower ball joint vertically, sets the chassis ride height.

We can install spindles with different pin offsets (distance from the lower ball joint) to change the race car ride height. If we change the spindle pin location, and keep our chassis ride height the same, we have then changed the lower control arm angle.

Spindle Height Effect on Camber Change

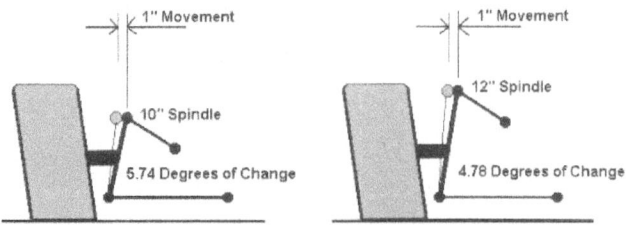

The Camber Change of the 12" Spindle is Nearly One Degree Less Than the 10" Spindle for the Same amount of Ball Joint Movement

The height of the spindle has an effect on the camber change, a very important design function we will learn more about later on. This illustration shows now many degrees of camber change results from different spindle heights. The taller spindle changes less.

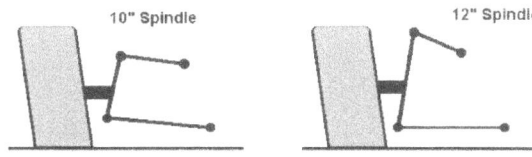

Effect of Using Shorter Spindle on Full Sized Stock Cars

Results

Longer Spindle has a higher Roll Center with better Camber Control and Less angle in the lower control arms helping to reduce the migration of the Roll Center after chassis Dive and Roll.

Shorter Spindle has a low Roll Center, with resulting poor Camber Control. The lower control arms have more angle which causes excessive migration of the Roll Center after chassis Dive and Roll.

Longer spindles can offer a better design choice for several reasons. First, like we illustrated before, the camber change is better for the same movement of the ball joint. But the taller spindle also creates more angle for the upper control arm and that also has a positive effect on camber change reducing it further.

In a front view, as we discussed with scrub radius, the king pin angle is a design variable. We necessarily need different king pin angle designs for the right and left spindles in a circle track car. This helps reduce the camber change that results from steering the wheels.

Just as caster causes camber change, so does king pin inclination. These two combine their effects to end up with a net camber change per degree of wheel steering movement. All of this is taken into account in the overall suspension design and you don't need to completely understand all of that at this point in time. Just know that a spindle can be designed with different king pin angles.

Steering Arm - Since the spindle is attached to the wheel, and the wheel has to turn, a steering arm must be attached to and be a part of the spindle. This arm extends either forward or rearward and is attached to the steering system through a tie rod. This rod "ties" the spindle to the steering box or rack.

The steering arm for each spindle can be placed at a predetermined height, at different lengths and with the end different distances from the centerline of the chassis. The ultimate position and length of the steering arm coincides with the overall design of the steering system.

This is all you need to know about steering arms at this point in time. We will get into more detail about steering systems in future Lessons as a part of RCT Level Two and RCT Level Three.

A race team can create a taller spindle just by installing a ball joint with a taller, or longer, shaft. The height of the spindle for design purposes is the measurement between the centers of rotation of the ball joints. This longer ball joint shaft raises the center of rotation creating a taller spindle.

Mixing Different Spindle Designs - What we need to stress here is something that is important to understand from a basic perspective. We do not want to mix and match spindle designs without completely understanding how we might be changing our steering system, our front geometry layout and other considerations and design functions.

In the past, with some forms of racing, chassis manufacturers and teams have substituted spindles of different dimensions in order to save weight. In doing this, they inadvertently cause serious negative changes to the front ends of the cars and the performance suffered. It was some time before they all figured this out and redesigned the cars to work with the new spindles.

The spindle can be any design, including this very simple one made out of steel tubing and using simple heim joints for ball joints. This spindle is made for a scaled down stock car that weights much less than a full sized stock car and has much less loading on it.

and height all need to coincide with the overall design of the steering system.

Formula race cars have specific designs for their spindles where they cannot be interchanged. It is mostly with stock car racing where this occurs and a race team must be careful not to make undesirable changes with components without understanding how those parts will affect the suspension functions.

Spindle Recap - So, to recap, a spindle is a solid part that attaches to the control arms through the use of ball joints, or heim joints. These terms are interchangeable.

The spindle is the connection between the chassis and the wheel. It is attached to the wheel through a spindle pin that carries bearings. The bearings connect the spindle to the wheel hub and run on bearing races attached to the wheel hub.

The spindle helps support the chassis. Since the springs are mounted to the lower control arm, and the lower control arm extends to the spindle, it rests on the ball joint that is attached to the spindle. The load of the chassis is extended from the spring mount through the ball joint and down through the wheel to the tire contact patch.

The upper and lower ball joints that are attached to the spindle are the three dimensional connection of the spindle to the control arms. The front view angle of the ball joints from a vertical line is the spindle inclination or king pin inclination, interchangeable terms.

The other function of the spindle is to provide a means to steer the car by the use of a steering arm as a part of the spindle. The steering arm lengths, position laterally

Exam - In The Context Of This Lesson:

The Spindle Connects Which Components?

1) The chassis to the wheel
2) The steering box or rack to the wheel
3) The spring to the wheel
4) The upper and lower ball joints
5) All of the above

Differences In The Following Affect Control Arm Angles

1) King Pin angles
2) Steering arm lengths
3) Spindle pin height
4) Wheel camber changes

Scrub Radius Affects The Following

1) Camber change
2) Caster change
3) Wheel Loading
4) Feedback through the steering system

Spindle Design Must Coincide With Which Other Components?

1) Control Arm Angles
2) Ride Height
3) Scrub Radius
4) Steering System Function
5) All of the above

Lesson Six – Control Arms

Control Arms are the connection between the chassis and the wheel. These are major components of the double A-arm suspension systems. These control arms can be found on either the front or rear suspension of the car. Most designs of control arms feature a long lower arm and a shorter upper arm. This is called the Short-Long arm suspension system.

Control arms make up a major part of the structure in a double A-arm type of suspension. Here we see the two arms, upper and lower, without the spindle attached. These are often called long arm/short arm suspensions because with most designs, the upper arm is shorter than the lower arm.

Control arms can be of a one-piece design, or be made up of different parts bolted together. The inner mounts will be bushings or heim joints and attached at two points to the chassis. The outer mounts are ball joints or heim joints (interchangeable terms) and those are mounted to a spindle.

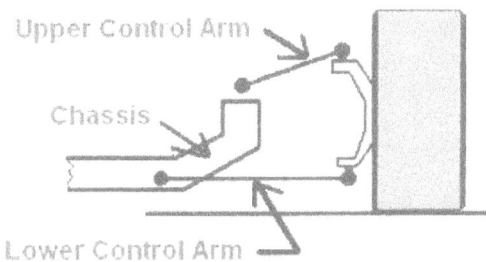

The control arms for a race car in a double A-arm suspension consist of an upper and lower arms that connect the spindle to the chassis. The outer mounts are called ball joints, or mono-ball, and the inner mounts are bushings, or heim joints. The angle of the arms is a critical component of race car suspension design.

Control arm angles are talked about in two separate views. From a front view, the upper control arm angle is usually with the chassis mount lower than the ball joint in stock cars and many prototype cars. The exception is formula cars such as Formula One. The angle of the upper control arm can be from five degrees to upwards of thirty degrees depending on the type of car and overall design goals.

The lower control arm, from a front view, can be either with the ball joint higher or lower than the chassis mount depending on the overall design goals. In either case, the angle is usually less than five degrees.

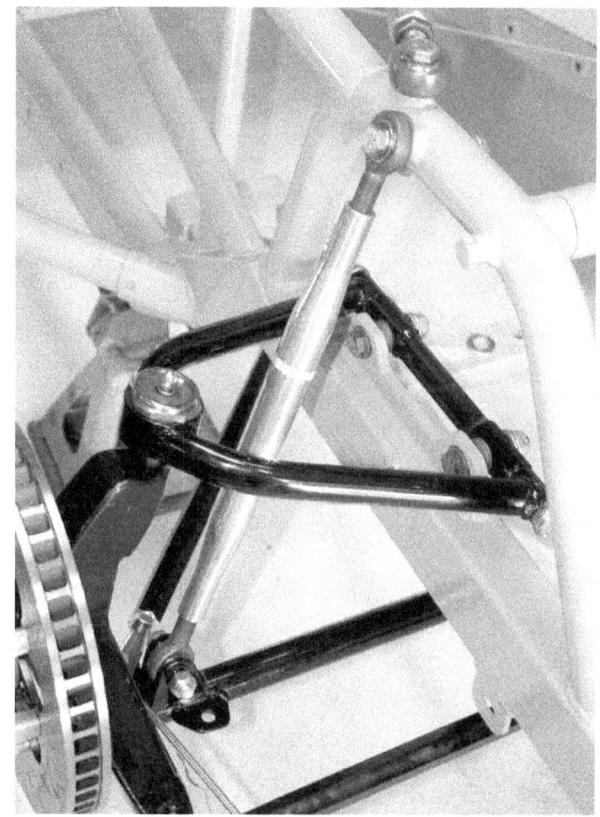

This is an example of a typical stock car, or tube frame road racing car, upper and lower control arms. The upper is one piece and adjustable for angle and camber by moving the inner mount through spacing or height of the control arm shaft. The silver link shown is a temporary link that replaces the coil-over shock and spring combination. Note that the springs that support the car are almost always attached to the lower control arm on almost every race car.

Control arm angles from a side view create anti or pro-dive, which is a function of braking force that we will get into later on. Just know that this effect exists and the side view control arm angles create the effect. The front view control arms positions and angles determine the suspension geometry components such as roll center and roll center movement, cambers and camber change, and anti and pro-dive.

This is an example of a double A-arm suspension on a prototype race car. The upper and lower arms consist of links attached to the spindle and chassis with mono-ball joints. The car is supported by springs mounted inboard on the transaxle housing and connected to the lower control arm/spindle with a push rod link.

In most cases with modern race cars, the spring that supports the chassis is mounted to the lower control arm. Where this spring is mounted determines the wheel rate and is the wheel rate is calculated using the motion ratio.

The lower control arm must be strong enough to support the forces generated by the spring through the amount of wheel travel that occurs on the race track.

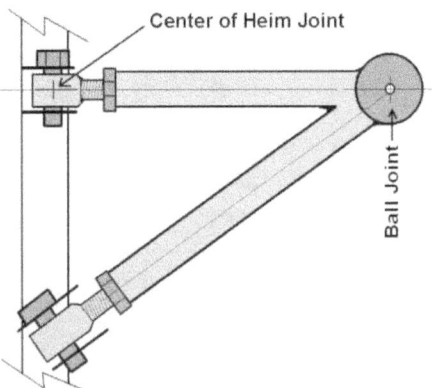

The upper control arm can consist of links mounted to the chassis with heim joints.

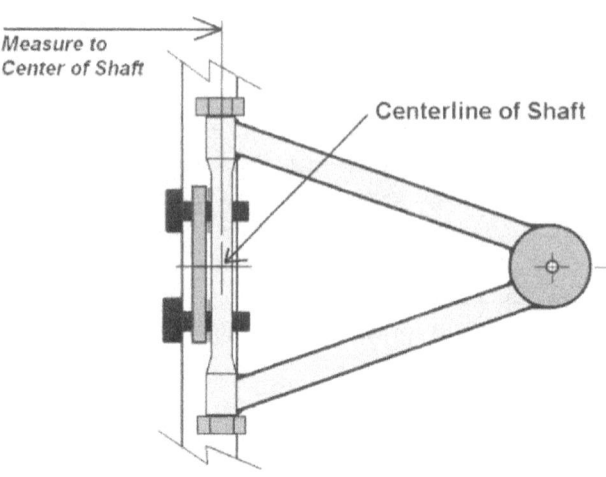

The most common upper control arm design for stock cars and many road racing cars is a one-piece design with a shaft attached to the chassis. The shaft can be moved in towards the centerline of the chassis to gain camber or out to remove camber. The shaft can be moved fore or aft to adjust the amount of caster.

Adjusting Upper Control Arm Locations

The upper control arms are usually adjustable for inner mounting heights and widths. If the control arm is solid, then the inner mount will be a shaft which can be spaced on the chassis mount in or out to make changes to the wheel camber.

The height of the upper control arm chassis mounts can be changed on any design and these changes are used to change the arm angles. The change in upper control arm angles produce changes in roll center location and migration. The upper control arm front view angles also affect camber change characteristics.

In most systems, the upper control arm can be moved fore and aft to create caster and caster split. If the upper control arm is a one-piece design, many of those are designed with slotted shafts where the team can loosen the attachment bolts and slide the arm fore or aft to make changes to the caster settings.

We can see the spacer plates positioned between the control arm shaft and the chassis mounting plate. The plate has slots and slugs with varying offset holes to move the shaft up or down to adjust the arm angle. Note too that the shaft is slotted to allow movement fore and aft to adjust the caster.

In the upper control arm mounts, we also usually see slots or slots with slugs so that the inner mounts can be moved up or down to change the front view angle of the control arm. The upper arm angle can also be changed by installing a ball joint with a longer or shorter stud. Some cars use mono-ball mounts where the ball can be spaced at different heights on the spindle mounting stud.

From a side view, the mounting bolts for the upper control arm can be different heights. This creates anti or pro-dive. For anti-dive, the front mount is placed higher than the rear mount.

Some upper control arms are adjustable for length using the same chassis mounting points and same ball joint heights. This adjustment is meant to speed up camber changes. Remember that when doing the camber change using adjustable length control arms that you are also changing the arm angle and with that the geometry of the front suspension.

The Lower Control Arms

The Lower control arms are usually less adjustable for front view angles than the upper arms. In side view, there can be changes to the angles to support anti and pro-dive designs.

The length of the lower control arm compared to the distance the spring is mounted from the ball joint determines the motion ratio for that control arm.

The lower control arm on a stock car for circle track or road racing mostly supports the car by being attached to a spring or torsion bar. Here we see a stock spring lower control arm where the spring is resting in a round bucket that is a part of the lower control arm. The outer end of the arm rests on the lower ball joint.

Motion ratio is: the difference in distance that the wheel moves comparted to the movement of the spring. On most systems, the wheel always moves more than the spring. On some formula and prototype systems, the spring could move more than the wheel, but this is usually not the case.

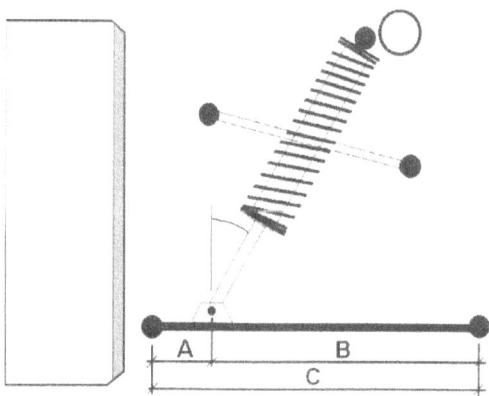

With the spring attached to the lower control arm, we can see where there is a motion ratio as the wheel moves. The spring always moves less than the wheel. The spring motion divided by the wheel motion is called the motion ratio. If the spring is mounted at an angle to the motion of the lower control arm, the spring motion is even less than the motion ratio.

As to anti and pro-dive, if the rear mount of the lower control arm is mounted higher than the front mount, we have anti-dive. If the rear mount is lower than the front mount, we have pro-dive. We will be discussing the anti and pro-dive functions in more detail in the next Lessons.

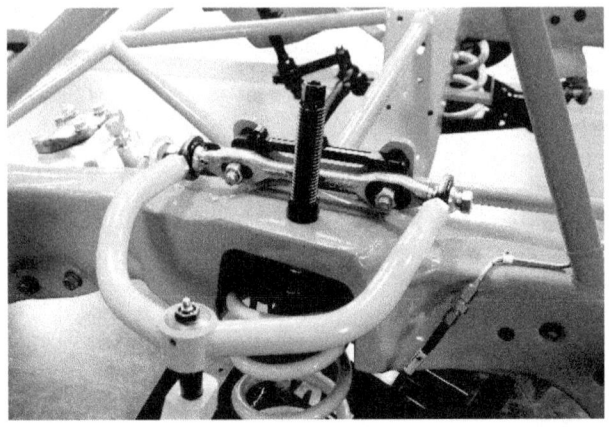

The angle of the control arm, upper or lower, is measured on a line passing through the centers of rotation of the ball joint and the inner mounts. This control arm is bent to avoid contact with the chassis as the wheel drops down, but the actual angle of the arm does not change.

Race cars with this type of upper chassis mount cannot make fine adjustments to the arm angle. The length of the arm, the cambers and caster settings are adjustable through the adjustable links that make up the arm. To change the arm angle, ball joints with longer or shorter shafts will need to be installed.

Exam - In The Context Of This Lesson:

Control Arms Connect Which Components?
1) The tie rd to the spindle
2) Chassis to the spindle
3) Wheel to he spindle
4) The braks to the spindle

Side View Angles Of The Control Arms Create?
1) Roll center locations
2) Wheel cambers
3) Anti or Pro-dive
4) Caster Settings

Changes To The Front View Angles Of The Control Arms Affect?
1) Roll center height
2) Roll center width
3) Camber change
4) All of the above

Which Suspension Component Helps Support The Car?
1) Upper control arm
2) Roll center height
3) Roll center width
4) The lower control arm

Using The Motion Ratio, We Can Determine?
1) The lower control arm angle
2) The ride height
3) Wheel rate
4) The upper control arm length

Lesson Seven – Anti and Pro Dive

Anti and Pro dive are design functions that either prevent the front suspension from diving when the driver is braking, or assists in diving when the driver is braking. To completely understand the Anti's and how they work, we need to understand what is happening to the car when we brake.

It is important to note that the Anti's will only function when there is braking force. The result of the braking force assists, and actually causes, the Anti's function.

There are various theories about how Anti's work in various teachings. In this school, we will deal with only one and that will deal with the actual forces in a practical way. This involves the force of deceleration and the torque created when we apply the brakes.

Sketch-01

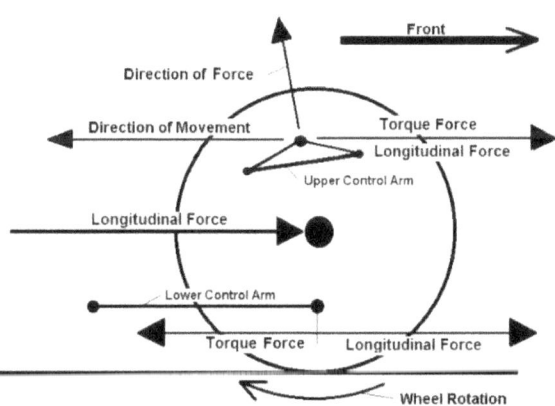

In this side-view sketch of the right front suspension, we see where the upper control arm is mounted at an angle. The rear arm bushing is lower than the front bushing. If we move the spindle up, or the chassis goes down, the ball joint will move in a direction that is to the rear, or back towards the rear of the car. Upon braking, the force of the brakes grabbing the rotor is trying to force the rotor and spindle that it is attached to in the direction of the rotation of the wheel.

Since these two motions are opposite, there is Anti-dive created. In addition, braking forces add to the torque forces and those two combine to create the Anti-dive effect. The chassis cannot, or is resisted from moving down by the forces acting to lift the chassis in opposition to diving.

On Braking – When we apply the brakes in a race car, there are several things going on. One, weight is transferred from the rear of the car to the front. This increase in weight on the front suspension causes the springs to be made to support that added weight, and so they compress.

When the spring compress, the front of the car goes lower to the track. Many teams don't desire this diving motion and so Anti-dive was created. It was found that by creating a side view angle with the upper and lower control arms, this tendency to dive was reduced or eliminated.

What Anti-dive does is lock up the suspension and force it not to move, or not to move as far. It does this by creating opposing forces to the motion of chassis dive. These forces are a combination of braking longitudinal forces and braking torque forces.

Looking at Sketch-01, we see where as the chassis moves down, or the spindle moves up, with the side-view angle in the upper and lower control arms, the ball joints will move in the opposite direction as the wheel rotation and the braking force and the chassis opposite to that.

Because there is a braking force trying to move the ball joints in the direction of the wheel rotation, and an opposite diving force trying move the ball joints in the opposite direction of the wheel rotation, these two forces oppose each other. This defines the torque aspect of Anti-dive.

Then too the longitudinal braking forces trying to drive the chassis forward puts a force on the ball joints as well as the chassis mounting points. When the control arms are angled, these forces act in a direction opposite of chassis dive, or upwards.

The opposing forces are the creation of Anti-dive. There is another effect that is the opposite of Anti-dive, and that is Pro-dive. If in our side-view angles of the upper and lower control arms, we mount the control arm angles in the opposite direction, we then can assist with chassis dive upon braking, if that is our desire.

Then the two forces are acting in the same direction and the chassis will dive quicker than if it had no Pro-dive angles in the control arms.

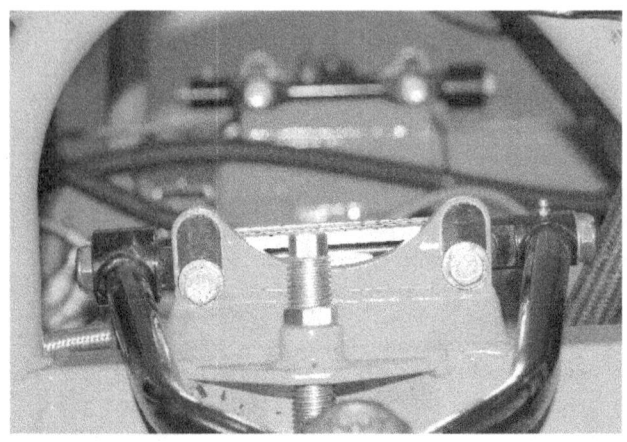

Here we see a right front control arm mount where the control arm shaft is not level and has some amount of Anti-dive. The front bushing is higher than the rear busing. If there were a corresponding angle in the lower control arm, then there would be Anti-dive in this system. To create more Anti-dive, we would raise the front mount by inserting a slug in the front slot that has less offset down, or offset up from the middle setting creating more sideview angle.

The only time we see this Pro-dive design used is on a circle track car, and on the left front suspension only. It is thought that it is desirable to cause the left front suspension to go quickly to a lower position by some setup designers.

It is not the intent of this school to comment, at this point, on the merits of certain setup routines and practices, so we won't. This school is created to teach the student about all of the chassis setup tools that are available and to understand the function of all of the components of a race car.

How Much Anti-Dive? – The amount of braking force that Anti-dive tries to control is directly in relation to the amount of braking that is taking place. Pressing lightly on the brakes cause less weight transfer to the front and therefore less dive. So, for light braking cars, there is less need for Anti-dive.

For heavy braking cars, there is obviously more need for Anti-dive and a greater amount of Anti-dive settings. Here is why we even need Anti-dive.

When a double A-arm suspension dives, it creates camber change in most cases. Camber change towards the negative will reduce the tire footprint and reduce grip. Less grip means the car does not turn in as well and it will take longer to slow the car.

For an average circle track car, or a road race car using a chassis similar to the circle track late model, a side view angle of the upper control arm of around 2 ½ degrees is considered moderate. Up to 5 degrees of angle in the uppers is considered acceptable.

This is a mounting plate for the right upper control arm. The rear offset slug has a ¼" offset down from center. The front slug has a 1/8" offset down from the center. This would create an 1/8" of difference in height, or about a 1.2 degrees of anti-dive angle. A half inch difference creates almost 5 degrees of anti-dive angle, about the maximum we would want to use normally.

The angle in the lower control arm is a definite part of Anti-dive and acts in the same way as the upper control arm. Its motion moves the lower ball joint much less than a similar angle would move the upper ball joint, and because the longitudinal and torque forces act in the opposite directions, there is much less effect with the lower control arm in Anti-dive than with the upper control arms. Angles from 2 to 4 degrees are acceptable for the lower control arm.

Exam - In The Context Of This Lesson:

Anti and Pro-dive Only Present When?
1) The car is accelerating
2) The car is braking
3) The wheels are moving up or down
4) The car is turning

With Anti-dive The Upper Ball Joint Moves?
1) To the front when the chassis dives
2) The In the same direction as the lower ball joint
3) In the same direction as the wheel rotation
4) To the rear when the chassis dives

The Amount Of Anti-dive Needed Is Dependent On?
1) How much weight is transferred
2) The length of the track
3) How hard the driver is braking
4) The length of the upper control arm

With Pro-dive The Upper Ball Joint Moves?
1) To the front when the chassis dives
2) The In the same direction as the lower ball joint
3) In the same direction as the wheel rotation
4) To the rear when the chassis dives
5) 1 and 3

The Primary Benefit Of Anti-dive Is?
1) Keep the front end up
2) Cause less weight transfer
3) Reduce or prevent camber change
4) Help the brakes work harder

Lesson Eight – Steering Systems

As described in Lesson Two, steering systems are on the front of a race car for a reason. We use the steering to generate an angle of attack to the tangent of the curve we are driving around. We will describe the different types of steering systems used on race cars and explain the function of the various parts of the steering system.

As a part of this Lesson, we will also explain how different classes of race cars use and setup their steering systems. It is not necessarily the same for all classes. We'll tell you why.

Types Of Steering Systems - There are basically three different types of steering systems used in racing today. They are:

Rack and Pinion Steering – In this system, a rack is moved by a gear when the steering wheel is turned. The rack ends are connected to the inside ends of the right and left tie rods that are connected to the spindle to turn the wheels.

The rack and pinion steering system is the most common steering in most race cars and today in most production cars. It is simple in design and uses just the center rack and housing with a pinion gear that moves the rack back and forth to steer the car. There are various design parameters for installing the correct rack that will be covered in future Lessons. A power assist unit (at the front of the gold rack) can be added to the rack to reduce the force needed to turn the steering wheel.

Drag Link Steering – This is an older system used on many American production automobiles and adapted to stock car racing. It consists of a steering box that moves a link. On the link are mounted the two tie rod ends, one left and one right. The link can be designed to be in front of the center of the spindle or to the rear. The front is called a Front Steer car and the one to the rear, a Rear Steer car. The Front Steer is the more popular of the two in today's racing.

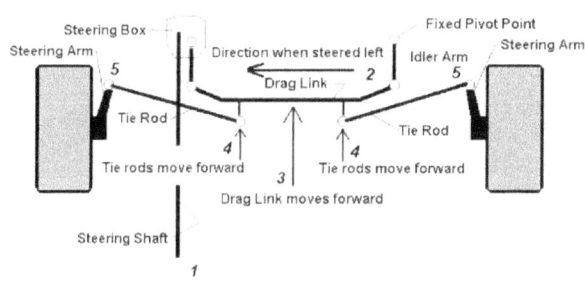

The Drag Link steering system is common on older production cars and some race cars. The layout is just as described, a steering box that when turned by the steering wheel, moves a drag link side to side. Attached to the drag link are the inner ends of the tie rods. As the link moves, so do the tie rods and wheels to steer the car.

Straight Axle Steering – On race cars, such as sprint cars, the front suspension is a straight axle design. To steer the car, a steering box is mounted just in front of the steering wheel and above the drivers legs. A shaft extends out of the right or left side of the car and a rod is mounted to an arm off the shaft. A tube link is then mounted to that arm and forward to a horizontal arm off the spindle on that side. The steering box moves the arm and tube link which moves the spindle arm turning the spindle. The other spindle is connected to the steered spindle by a cross link attached to steering arms that are a part of the spindles. This is a simple but effective steering system that has its flaws.

33

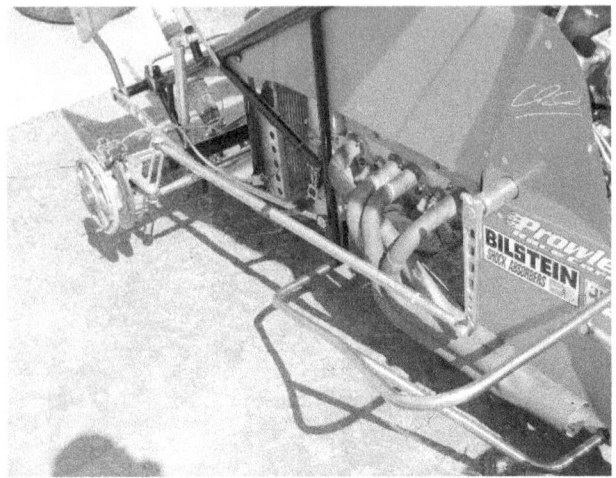

The steering on a straight axle is built differently than either the rack and pinion or the drag link systems. This unit uses a gear box located just forward of the steering wheel. This box turns a shaft that protrudes out of either side of the body. It has an arm at its end that moves a steering rod going forward to the spindle and a steering arm on the spindle that is parallel to the straight axle, as shown here. The steering arm also has a ninety degree arm connected to a tie rod that steers the other wheel.

Toe Defined - Toe is when one wheel is pointed in a different direction than the other at static ride height and with the wheel pointed straight ahead. We can have toe on both the front and rear wheels. Toe is used to stabilize the directional attitude of a vehicle and some production vehicles are designed with toe-in and some with toe-out in the front wheels.

To further explain what toe is, here is an example. If you could measure the width between the two tire centers of the tire tread at the front and at the rear of those tires, with the wheel pointed straight ahead, for toe-out, the front measurement would be more than the rear. Toe-in would be the opposite. In racing, we have various tools to measure toe.

Toe-out is most common for race cars for the front wheels. We change the toe in our race cars by lengthening or shortening the tie rods, which are designs to do this easily. Most tie rods have opposing direction threads on opposite ends of the rod end that attaches to the steering rack or drag link. When you turn the tie rod after loosening the jam nuts, the heim joints, or tie rod ends, move in or out together.

On many formula and prototype cars, the front wheels are set with toe-out and the rear wheels are set with toe-in. The toe can change as we negotiate turns and this change is caused by the suspension movement and the motion of steering the wheels. We'll talk more about those changes and why they happen later on.

When we measure the toe in our race cars, we can use toe plates that rest against the tires sidewalls, or we can use strings run parallel to the centerline of the car. There are also toe bars that offer a way to measure to the sidewall by first placing the bar in front, measuring an offset, and then to the rear to see how much more offset there is that is the toe amount.

We like to think of toe as the difference in the front and rear widths at the center of the tread. But if we measure at the sidewall, at the edge of toe plates, or even at the rim, which some teams do with the string method, we are not reading the toe amount at the tire tread, we are seeing something less.

This is because the width of where we are measuring is less than the diameter of the tire. If our measurement is less, then we are reading a number that is less than the real toe thought of as the difference in width of the centers of the tires at the contact patch.

To correct our numbers, we could divide the tire diameter by the width of the points we used to measure toe and then multiply that number times the toe number we get. For example, if we are measuring points 20 inches apart and our tire diameter were 27 inches, then dividing 27 by 20 gives us 1.35. If we measure our toe at 1/8", which is 0.125", then when we multiply 1.35 times 0.125, we get 0.169, or closer to 3/16" of toe.

And remember this too about toe, when we make changes to our wheel cambers, the movement of the upper control arm in or out does move the end of the steering arm, which changes the toe measurement. You must recheck and reset the toe after making changes to the cambers of the front wheels.

Steering Quickness - When we steer our cars, we need to have a steering ratio that is quick enough so that we don't need to take our hands off the steering wheel. And it should not be too quick so as to move the steering excessively with small movements of the steering wheel.

In stock classes where the teams are required to use stock steering boxes or racks, racing suppliers offer steering quickeners, which are a small geared box that speeds up the turning of the steering shaft.

Ratios of from 1.5/1.0 to 2.0/1.0 are offered. That means for a 2.0/1.0 ratio, if you turn the steering wheel 1.0 revolutions with the quickener installed, it would be like turning the steering wheel two times without the quickener. Now the driver doesn't need to cross over with the hands to turn far enough to steer the car.

There are mechanical effects that can cause the wheels to steer in directions other than what we intend to have happen. Those are:

Ackermann – When one wheel steers more than the other. When the wheels gain toe, it is called Ackermann. When the wheel lose toe, it is called Reverse Ackermann. We will cover the causes of Ackermann and how to fix it in Lesson 9.

Bump Steer – When the wheel steers due to vertical wheel movement, we have bump steer. Sometimes a very small amount of bump steer is desired, but normally we want near zero bump steer. We will cover what causes bump steer and how to measure it in Lesson 8.

How Much To Steer? - There is an optimum amount we need to steer. It is just beyond how much it would take to steer the car slowly through the turns. If our driver drove around through the turns at a slow speed, the amount of steering required to roll around the driving line would be nearly the same as when driving that same line at high speeds, but a little more.

If the driver finds that they need to steer a greater amount, then the car has a handling balance issue. This means that the front tires have less Grip than they need to keep up with the Grip of the rear tires. Changes to the setup must be made to bring the steering back to normal steering angles.

In dirt racing, many of the cars have a great amount of rear steer where the rear tires steer out away from the turn. This rear steer is so pronounced that the front wheels must be steered to the right to compensate and to keep the front tires tracking along the correct line. This is unique and very different than anything we see with racing on asphalt.

Steering As A Grip Generator - Steering creates Grip. When we steer the car, we are creating an angle of attack that produces a force we need to counter the lateral forces trying to push the car off the turns. The steering system is therefore a variable Grip generator. The more we steer, the more Grip we produce, up to a point.

As the angle of attack increases with increased steering angle, the Grip increases until we have gone too far. Then at that point, the tire can no longer hold to the track surface and it slides. The level of Grip the tire is producing drops off dramatically at this point and the front of the car will lose most of its Grip and slide.

This is why a car that is unbalanced and where the setup is tight (car is understeering), can still be made to turn and make it through the turns. The excess steering needed to do this forces the front tires to work harder than they would if the car were more neutral in handling balance.

Exam - In The Context Of This Lesson:

The Most Common Types of Common Steering Systems Is?

1) Tie Rod System
2) Caster System
3) Rack and Pinion
4) Rear Steer System

How Much Do We Need To Steer?

1) One full turn of the steering wheel
2) Till the rear stops sliding
3) Enough to track just inside the tangent of the arc we are driving
4) 1.5 times what we would on the street

Steering Quickeners Do What In Our Steering System?

1) Prevent excess toe out
2) Make the lap faster
3) Increase turn speeds
4) Make the car steer more with less steering input

Bump Steer Helps The Car Do What?

1) Use less camber
2) Become less stable
3) Increase turn speeds
4) Ride out bumps better

Ackermann Is?

1) When the inside wheel turns less than the outside wheel
2) When the car loses toe from steering input
3) When the car gains toe from steering input
4) A way to improve the steering system

Steering Generates Grip By What Method?

1) The creation of Ackermann
2) By reducing bump steer
3) Through increasing the steering quickness
4) By giving the tires an angle of attack

Lesson Nine – Bump Steer, Causes and Cures

When a wheel moves vertically, we say it bumps. This probably goes back to the day when our roads had a lot of pot holes and bumps. In that situation, if the car ran over a bump, automotive designers did not want the car to steer as the wheel moved over the bump. Thus came the term, "bump steer".

It is also true of our race cars that we don't want or need bump steer for the most part. In some more involved suspension designs, some amount of bump steer may be desirable, but here we will stick to the basics and say that bump steer is a bad thing.

Attaining zero bump steer is nearly impossible because of the complexities in our steering system, but we can design for near-zero bump steer that is so small it won't be noticed by the driver.

There are a few very complicated and sophisticated machines that measure bumps steer on race cars under chassis loading situations. We can come to understand all we need to know about bump steer by learning just a few important facts about bump steer.

What Creates Bump Steer? - The angles of the upper and lower control arms, meaning a line extending through the center of rotation of the ball joints and inner mounts of each arm, intersect at a point we call the Instant Center. In order to have near zero bump steer, the tie rods on each side need to be pointing towards the instant center for its side. This is one of two criteria for near zero bump steer.

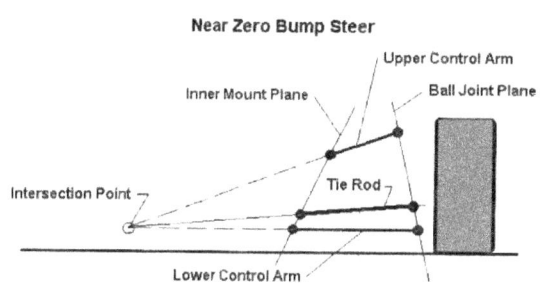

The basic criterial for having near zero bump steer is shown in this sketch. The tie rod is pointing at the instant center formed by the extension of the upper and lower control arms. And, the tie rod is the correct length, being the distance of its lines intersection with the two planes, one through the upper and lower chassis mounts, and the other plane through the ball joints.

The other thing we need is for the tie rod to be a specific length. That length must be equal to the distance formed by a line extending through the centers of rotation of the tie rod ends, and that lines intersection with two planes, one formed by the upper and lower ball joints and being parallel to the wheel, and two a plane that passes through the inner upper and lower chassis mounts. This can get a little complicated because it may be hard to visualize these two planes.

And, the tie rod does not necessarily need to physically fall within those two planes, it just needs to be that length between the points of intersection with each plane.

If we have the tie rod pointing to the instant center intersection of the upper and lower control arms and it is the right length, then we will have near zero bumps steer.

What Creates Bump Steer – When the tie rod is not aligned with the instant center and/or the length is wrong for the system, we have bump steer. As the wheel moves vertically, the wheel will either steer left or right. We will refer to the direction from a driver's perspective and for a front steer car only, in this discussion. By front steer, we mean that the tie rods are in front of the ball joints.

If the tie rod was pointed so the tie rod line passes below the instant center, then the wheel will bump in (towards the centerline of the car) as the wheel travels up and bump out when the wheel travels down. If the tie rod line passes above the instant center, then we will have bump out as the wheel travels up and bump in when the wheel travels down.

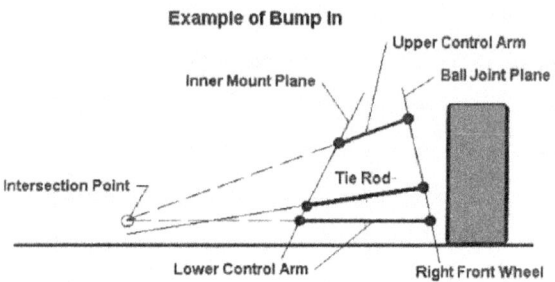

If the tie rod were pointed below the instant center, then when the wheel bumped up, the wheel would steer in, or left in this case. Other suspension design features can interfere with the bump steer layout and cause bump steer. This is caused by suspension movement that changes the alignment of the tie rod.

If the tie rod were too short, we would have bump steer in when the wheel travels in both directions from the static ride height position. If it were too long, then the wheel would bump out as the wheel traveled in both directions from ride height.

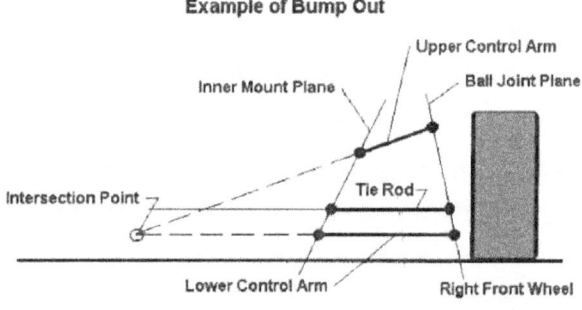

If the tie rod were pointed above the instant center, then when the wheel bumped up, the wheel would steer out, or to the right in this case. Changes to correct bump steer include movement of the ends of the tie rod either up or down, until it is aligned with the instant center.

We can easily check the race car for bumps steer by using a bump steer gauge. This measures differential movement of the sides of the plate bolted to the wheel hub as the spindle is jacked up from a preset position. This is a two dial gauge. There is also a one dial gauge where a roller replaces the one dial and the operator only reads the one dial. On the left front wheel, if the front dial was showing movement away and the rear dial showing movement to the operator, then this would be bump in.

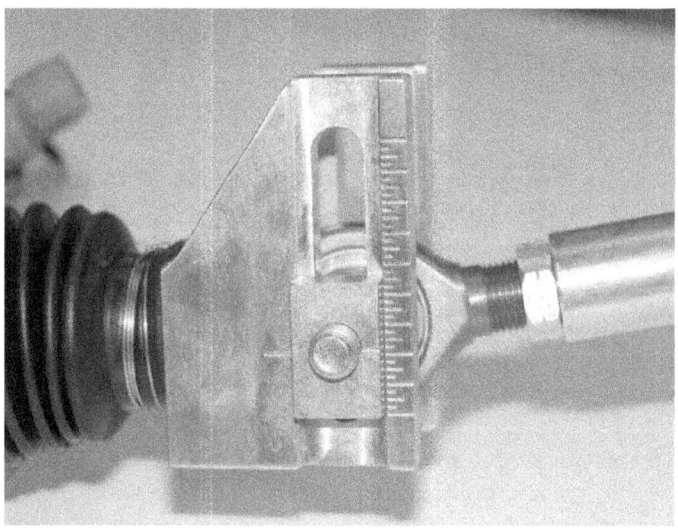

To correct bump steer on this rack and pinion steering system, we need only move the inner tie rod end up or down in this easy to read fixture. Drag link systems use spacers to move the drag link up or down, or offset slugs where the tie rod mounts to the drag link.

If we start making changes to the control arm angles, or other changes that would affect components needed for near zero bump steer, then after making those changes, we would need to re-adjust our bump steer settings.

For example, if we install extended lower ball joints to take angle out of the lower control arms, we change the angle of the lower arm and move the instant center height. After making that change, the tie rod may not now intersect with the instant center and we will have introduced bump steer. When making changes to the arm angles, we need to re-align the tie rod so that it stays pointed towards the instant center.

Another example is when we change our Anti-dive settings. In both dirt and asphalt racing, anti and pro-dive is used in various degrees. Introducing these effects cause changes to our bump steer. This is because with anti-dive for example, when the wheel travels up, the upper ball joint moves towards the rear of the car and this rotates the spindle from a right side view, counterclockwise. This rotation moves the outer tie rod end upwards and changes the angle of the tie rod. Now it no longer points towards the instant center.

Where we had near zero bump steer before with no anti-dive, we now have bump steer when the right front wheel travels up. With pro-dive, we see a similar affect, the tie rod end moves down with vertical travel and again the tie rod is miss-aligned with the instant center.

For checking tie rod length on a rack and pinion system, you can usually look to see if the inner end of the tie rod is lined up with the inner mount for the lower control arm. The final check is the bump steer gauge.

On this car, a spacer (red colored aluminum piece) has been placed between the end of the tie rod and the steering arm. In this case, the tie rod was pointing below the instant center by quite a bit. This is a large spacer.

Steering Affects Bump Steer – When we steer our front wheels, we change the angles of our tie rods due to caster, camber and degree of spindle on both sides. The tie rod ends travel in an arc that is not parallel to the ground. This changes the outer tie rod end height and therefore the B/S. It is for this reason that we recommend doing your B/S with the wheels both straight ahead and then again with the wheels turned equal to mid-turn steering at the track you will run.

Exam - In The Context Of This Lesson:

The Most Common Types of Steering Systems Is?

1) Tie Rod System

2) Caster System

3) Rack and Pinion

4) Rear Steer System

Which Is One Of The Two Elements Of Bumps Steer?

1) Tie rod angle

2) Tie rod length

3) Control arm angles

4) Control arm lengths

5) 1 and 2

Changes To Which Will Affect Bump Steer?

1) Tie rod angles

2) Tie rod lengths

3) Anti-dive

4) Caster settings

5) All of the above

Lesson Ten – Ackermann Steering

In every steering system, there is a design effect called Ackermann. It is the difference in turning rate for the inside and outside wheels. It is named for the man who came up with the effect and intended it to be used to track the front wheels of very early automobiles through gravel driveways without causing ruts. In today's race cars, there is very little need for the Ackermann steering effect.

Ackermann can be very destructive to a setup. When Ackermann effect is present in the steering system, the two wheels and tires will not track correctly and want to go in different directions. In this lesson we will explain what Ackermann is, how it is created and how to eliminate it.

Every steering system has within its geometry the means for a layout that produces Ackermann. For true Ackermann, the inside wheel (meaning to the inside of the turn) will always turn more per degree of steering input, than the outside wheel.

When this happens, the toe of the front wheels is increased. For production vehicles that may run toe-in, there is a loss of toe-in and the front wheels may well go into a toe-out condition. Reverse-Ackermann is the opposite condition, where the outside wheel turns more than the inside wheel. This produces reduced toe.

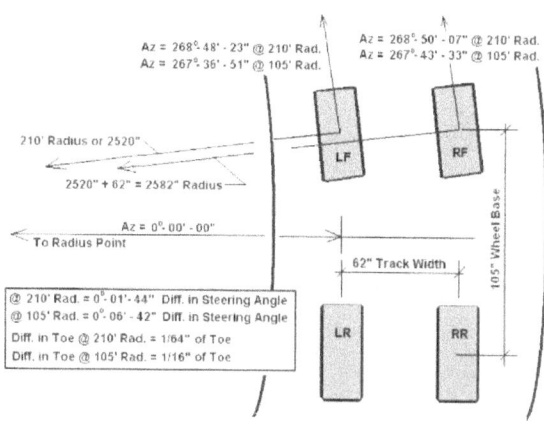

This calculation shows how little gain in toe we need in order for the wheels to track in their own path around the arcs that are slightly different from the inside wheels to the outside wheels. One degree of Ackermann for a typical race car going around an average radius turn is one half inch of toe. In the Lesson on Alignment, we will explain how that is a lot of toe out.

For most race cars that run toe-out, this would be increase. For average diameter tires run on race cars, one degree of Ackermann, which is a one degree of difference in steering direction of the front wheels, equals a half inch (0.500") added of toe-out. This is excessive.

For normal radius turns, we need fractions of a degree of Ackermann in our race car steering systems. Calculations tell us that for a 200 foot radius turn, we need about 1/64", or 0.015" of added toe. For a turn with a shorter radius of 100 feet, the added toe would be about 1/16", or 0.0625". If we set our car with 1/8" (0.125") of toe for a 1/3 or ½ mile track normally, if we go to a very short ¼ mile track, we only need to add 1/16" of static toe to have the needed 1/16" of Ackermann toe for that short radius turn. Then we don't need to redesign out steering system to produce the 1/16" of Ackermann.

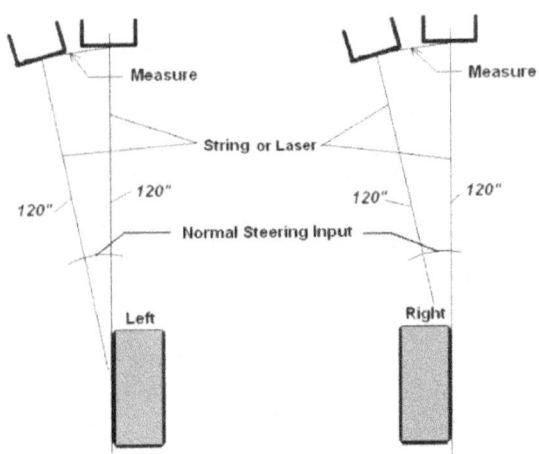
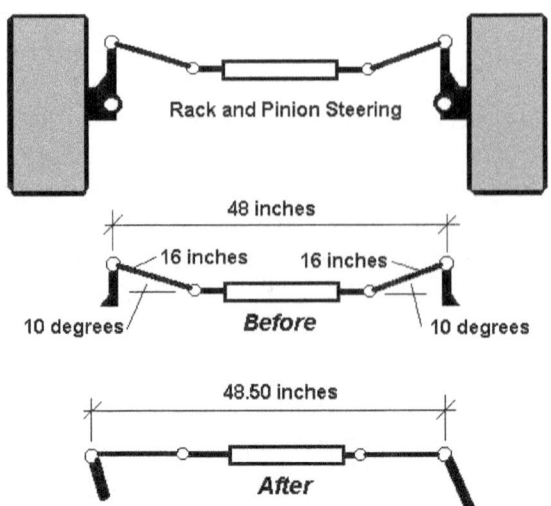

This sketch shows now we test for Ackermann. We set the wheels straight ahead and mark where they point at a distance from the center of the hub, in this case, ten feet or 120 inches. Then we turn the steering wheel usually the same amount as when we race the car. Two more marks are made where the wheels point after turning the wheel. If the measurement between those two sets of marks is different, then we have Ackermann if we have gained toe, or Reverse Ackermann if we lost toe.

How Steering Systems Produce Ackermann - Every steering system can be designed to create Ackermann, whether we want that effect or not. So, we have to be aware that when we assemble our car, or when someone else has assembled it before us, it might just have the tendency to produce unwanted Ackermann.

If we look at our steering system from a top view, we begin to see how Ackermann is produced. For a front steer, rack and pinion system, as the wheels are turned, the outer ends of the tie rods move backwards in relation to the rack. If the steering arms were mounted straight ahead on the spindle, and most of them are, and the tie rods were mounted so that the outer ends were ahead or forward of the inner tie rod ends, then as the wheel were steered, the outer tie rod ends would move back, the tie rods would straighten out, and that motion would push the outer ends apart. This produces Ackermann and increases the toe-out.

For a rack and pinion system, we can regulate and adjust the system for Ackermann by moving the rack forward or back. This does what we need for setting the Ackermann for turning both left and right as in road racing and when running on a dirt circle track. In this example, both steering arms are the same length.

The opposite would happen if the tie rods were mounted inline. Then as the outer tie rods ends moved back and the tie rods gained top view angle from originally being inline, the tie rod ends at the spindle would become closer together. This would cause Reverse Ackermann and we would lose toe.

The drag link steering systems work in a similar way. The big difference is that the steering arms are designed to be mounted at a top view angle on the spindle so that the outer tie rod end is closer to the centerline of the car than the lower ball joint. This angle in and of itself would produce Reverse Ackermann. When this design is used with a system that produces Ackermann, the two oppose each other and we get, as a result, near zero Ackermann.

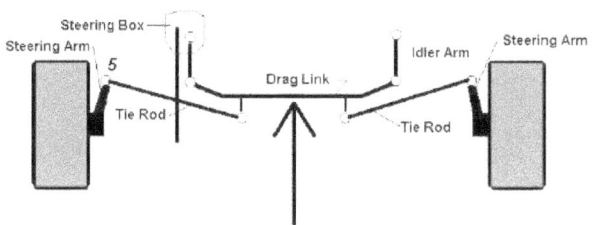

Be sure to look over your system to make sure it does not produce excess Ackermann. There is a simple string test you can do to check your Ackermann. If you have one, you can also use a laser system to do this check, but by all means, check it before you go racing. We'll tell you more about how to do this check in the Advanced RCT course.

For a drag link system, the adjustment is similar to when we adjusted the rack and pinion. Instead of moving the rack forward and aft, we can regulate and adjust the system for Ackermann by moving the drag link forward or back. This also does what we need for setting the Ackermann for turning both left and right as in road racing and when running on a dirt circle track. As in the rack example, both steering arms are the same length.

As a little history lesson, most of the stock cars produced in the eastern U.S. in the 1990's were drag link steering designs which used top view angled steering arms. When the stock cars of the mid-west, that used mainly rack and pinion steering systems, went to lighter spindles, some manufacturers in the east switched spindles and mounted the lighter rack spindles, with top view straight ahead steering arms, on their cars to save weight. The two steering systems were not compatible and the drag link east cars with the mid-west rack spindles were producing a lot of Ackermann and not turning too well.

If we only turn one way, we can adjust the length of one steering arm to tune the Ackermann amount. On most circle track cars, the left spindle has a slot for moving the tie rod end fore or aft to quicken or slow the turning of that wheel. For road racing cars, the steering arms on both spindles would need to be slotted and the adjustments made the same for both sides in order for the Ackermann to be the same for turning left or right.

Exam - In The Context Of This Lesson:

Ackermann Is Defined As?

1) The loss of front wheel toe

2) When the wheels bump steer out

3) A gain in toe when the wheels are steered

4) A loss of toe when the wheels are steered

Which Steering System Can Produce Ackermann?

1) Straight axle steering

2) Rack and Pinion steering

3) Drag link steering

4) All of the above

How Much Ackermann Is Needed For A 200 Foot Radius Turn?

1) One degree of Ackermann

2) A half degree of Ackermann

3) 1/64 inch of added toe

4) 1/16 inch of added toe

Rack and Pinion Steering System Require?

1) Top view straight ahead steering arms

2) Top view angled steering arms

3) A drag link attached to the tie rods

4) A shorter left steering arm

For Road Racing Cars, We Adjust Ackermann By?

1) Shortening the left steering arm only

2) Shortening the right steering arm only

3) Moving the rack or drag link forward or backwards

4) 1 and 2

Lesson Eleven – Front End Wheel Alignment

One of the most critical elements of race car setup is alignment of the wheels. The direction they are pointed at any point on the race track helps determine how well the car will handle and what the driver feels. We start with the alignment of the wheels when the car is at static (not moving) ride height and with the front wheels steered straight ahead.

We covered the concept of wheel toe in Lesson Nine because we needed to explore the Ackermann effect that can change the toe settings in our race cars. So, that being done, we'll get on to what the basics of alignment are.

Basic Alignment involves setting the suspension so that the tire contact patches line up when at static ride height and at mid-turn attitude. Alignment is different for circle track cars and road racing cars. For circle track alignment, we align the right side tire contact patches when the track width is different at each end of the car. For road racing, we align the contact patches to be equidistant from the centerline of the chassis. We have equipment available that can accurately measure the wheel alignment using lasers.

The most basic areas of study for alignment are:

- **Overall Wheel Alignment** – The alignment of each wheel in relation to the centerline of the car, or to a datum line we establish for our race car.

- **Tire Pair Alignment** – The alignment of the right side tires and the alignment of the left side tires to each other.

- **Rear Alignment** – The alignment of the rear in both static and dynamic conditions

- **Tire Pair Alignment Dynamically** – The way the side tire pairs align after the car has dived and rolled

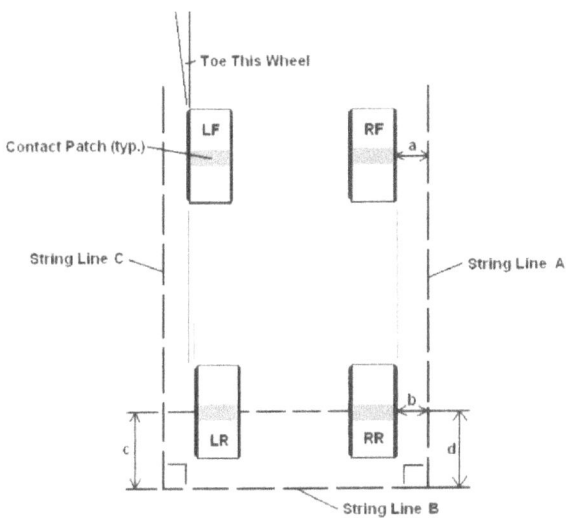

If we were to look down on the layout for alignment, it would look something like this. We concentrate on the tire contact patches because we often have camber in our wheels and aligning the tires at the hub height will not tell us if the tire contract patches are in alignment for cars with a-symmetrical cambers. This shows how a race team can use either strings or lasers to form a box around the outside of the car in order to measure for alignment.

Basic Understanding Of Alignment Components - We need to know how our wheels are aligned both when driving straight ahead, and when we are driving through the turns. The motion of the chassis can change the alignment we see at static ride height and if it does, we need to be able to measure that and adjust for the change.

When we discuss alignment, we are talking about the alignment of the tire contact patches. We often measure for alignment using the outside of the tire or rims at hub height. If the wheels have camber, then the contract patch is somewhere other than where the center of the tread is at hub height.

So, when we measure for alignment at hub height, we need to compensate our measurements so that we are aligning the contact patches, or the center of the tire tread at the ground.

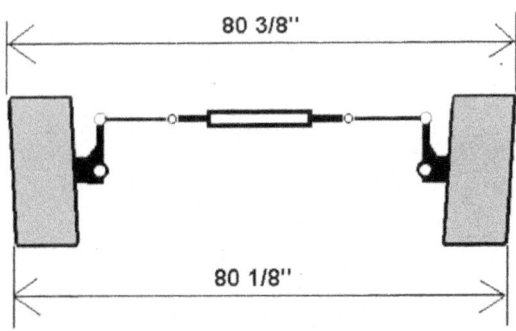

Toe is, as covered in previous Lessons, the aligning of the front, or rear, tires so that they are either closer together at the front than the back of the tire (Toe In), or the other way around (Toe Out). For most racing applications, the front wheels are Toed Out and the rear wheels are Toed In. Alignment also involves toe settings, but we set zero toe when we are aligning the four tires. Then after the four have been positioned correctly, we can then re-set the toe.

The front tire alignment is associated mainly with toe and Ackermann. We have studied both of these. But when we are checking alignment in relation to the rear tires, we need to eliminate the toe temporarily. The toe can be reset after we are finished checking and adjusting the alignment.

What we want as a goal for alignment is to make sure first of all that the rear tires are tracking in the same path as the front tires. This is the same process that is used on production cars at the alignment shop.

The alignment goals are a little different between a circle track car that only turns one direction and a road racing car. If the track widths (width of the sets of tires at each end of the car measured from center of contact patch to center of contact patch) on the front and rear of the car are different, then here are the requirements.

The use of lasers has become commonplace for checking and setting alignment and toe. The use of lasers can speed up the process of alignment and also make your work much more accurate than when using strings, which is an older method of alignment.

For circle track racing, we mostly line up the outside (right side on a left turning car) contact patches at static ride height. The rear track width is usually different and less than the front, so the inside tire contract patch will track to the right the front tire contact patch for a left turning car.

For road racing cars, we split the difference and align both ends so that the tire contact patches are equidistant from the centerline for each pair of tires on each end of the car. If the cambers are symmetrical at the front and rear, then we can align the car measuring at the hub height.

We can see where a laser line can be easily read on a ruler and be projected far beyond the ends of the car.

Alignment Can Change From Chassis Movement - The front track width can change due to chassis dive and roll. If the chassis moves enough to change the track width more than an $1/8^{th}$ of an inch, then you might consider compensating for the movement when you are aligning the tire contact patches.

The best way to do this is to travel the chassis as much as it does going through the turns and then align the wheels. Many companies now offer pull-down rigs that will pull the chassis to replicate the shock travels that exist in the turns. This does two things, one it places the tire contact patches where they truly are when the chassis travels. We can then know that we are truly aligning the contact patches to be correct, or where we want them to be, at mid-turn.

When checking for alignment, we can install solid links in place of the shocks or coil-overs so that the chassis doesn't move and so that we can raise the car in order to work under it. If we are using a pulldown rig to check alignment at dynamic ride height, we install the springs and shocks that will be used when racing.

The other important thing a pull-down rig does is load the suspension just as it is loaded on the race track. Then any deflection of the chassis components that might affect alignment are properly deflected and we can then know our alignment is true.

The front alignment can change when we steer the wheels and when the wheels move vertically. The rear alignment can change, and often does, when the car dives and rolls. For cars with a straight axle rear suspension, the links that position the rear fore and aft as well as laterally, can move the rear tires side to side as the chassis moves.

When the straight axle rear moves side to side, it can also change its direction of alignment. This is called rear steer, because the rear tires are actually steering, just like the front tires. This action can be very detrimental, or helpful, to the setup. It all depends on what our goals are.

Exam - In The Context Of This Lesson:

Alignment Is?

1) Toe-out from bump

2) The direction the wheels are pointed

3) Amount wheels steer

4) Toe-in from bump

Alignment Can Change Due To?

1) Bump Steer

2) Setup changes

3) Chassis movement

4) Ackermann steering

5) All of the above

The Primary Goal For Alignment Is?

1) To balance the setup

2) To make the car neutral in handling

3) To make the tires track inline

4) To keep the tires from wearing

For Circle Track Racing, We Alignment The?

1) Front tires

2) Left side tires

3) Right side tires

4) Rear tires

For Road Racing, We Alignment The?

1) Front tires

2) Left side tires

3) Right side tires

4) Rear tires

5) Front or Rear tires equal distance from centerline

Race Car Technology – Level One
Lesson Twelve – Sway / Anti-Roll Bars

In race car setup, we control chassis roll by using what are called sway bars or anti-roll bars (the terms are interchangeable), the later better describing what they do. Not all race cars use sway bars. Dirt cars mostly don't use sway bars. The actual roll angle a chassis will roll to is variable and determined by:

- *Spring Stiffness* – Simply said, the stiffer the spring package is, the less roll we will see.

- *Magnitude of G-force* – The greater the lateral force through the turns, the greater the roll angle, in most cases.

- *Track Width and Height of Center of Gravity* – Roll angle is calculate knowing the Center of Gravity height, the track width, the lateral force and the weights of the race car. A wider track width reduces roll angle and a higher CG increases roll angle.

- *Track Banking Angle* – The greater the banking angle, for the same G-force, the less roll angle

- *Anti-Roll Mechanism Stiffness* – The stiffer the anti-roll mechanism, the less roll angle

The anti-roll, anti-sway, or roll bar systems in a race car are designed to restrict and reduce chassis roll as the car goes through the turns. In this way, the ride springs can be softer for ride considerations while the sway mechanism helps control the chassis roll. Excess roll can affect our cambers and other settings and is not considered beneficial.

These systems can be constructed in many different ways, like this example of a prototype rear suspension that uses links (white bars) that connect the spring rocker arms to the sway bar and blades at the rear behind the transaxle.

A sway bar is intended to limit the roll of a chassis so that the springs that are installed don't need to be overly stiff in order to resist roll. Excess roll can be very detrimental to a race car because it causes more suspension movement which can affect alignment, camber settings and weight distribution.

The actual sway bar mechanism can be designed in many ways. In all designs, the bar or material used in the mechanism, must bend or twist as the chassis rolls, not in dive. The vertical movement of the chassis is resisted by the springs. The lateral, or roll, movement of the chassis is resisted by both the springs and the sway bar mechanism.

In a modern stock car, the anti-roll system consists of a roll bar linked to the lower control arms by way of an arm that is attached to the control arm. This bar has an adjustable end on the part of the arm that attaches to the bar. This is called a three piece sway bar, the actual straight bar and two arms.

What Determines The Sway Bar Stiffness? - Sway bar stiffness can be determined. Here are the factors that determine the stiffness of a system.

- *Sway bar diameter* – The outer diameter of the bar, and if hollow, the wall thickness of the bar both help determine the bar stiffness. The larger the diameter and the thicker the wall, the stiffer the bar will be.

- *Sway bar length* – The length of the portion of the bar that will twist helps determine the bar stiffness. The longer the bar, the softer or less stiff the bar will be.

- *Sway system material hardness* – The material the twisting and/or bending parts are made of help determine stiffness. Of course, the harder the material, the stiffer the system will be.
- *Length of Arms and/or Blades* – The length of the arms and/or blades relate to the stiffness of the system. The longer the arm or blade, the less stiff the system will be. This is because of the leverage affect.
- *Angle of Blade to Motion* – In most prototype systems the blades can be turned in relation to the direction of motion of the links to the suspension parts. If flat to the link, the system will be the softest in roll resistance. If turned more edge wise, or parallel, to the link, they will bend less and offer more resistance to roll.
- *Motion Ratio of System* – The motion ratio of all these systems relate to the overall stiffness. On stock cars, if the sway bar arms are mounted closer to the lower control arm chassis mounts, there will be less movement of the arm and less roll resistance.

In the prototype, the farther away from a one-to-one motion ratio (the link moving less than the wheel), the less roll resistance. In more of a one-to-one ratio where the link moves nearly as much as the wheel, the roll resistance will be greater.

Stock Car Sway Bars - The bar in a stock car may be made as a one-piece type where a straight portion extends across the frame rails, and then is bent to angle back to the be mounted under the lower control arms. Another design utilizes a straight bar with serrations on the ends. Separate arms are then clamped onto the ends of the sway bar and then extend back to be mounted under the lower control arms.

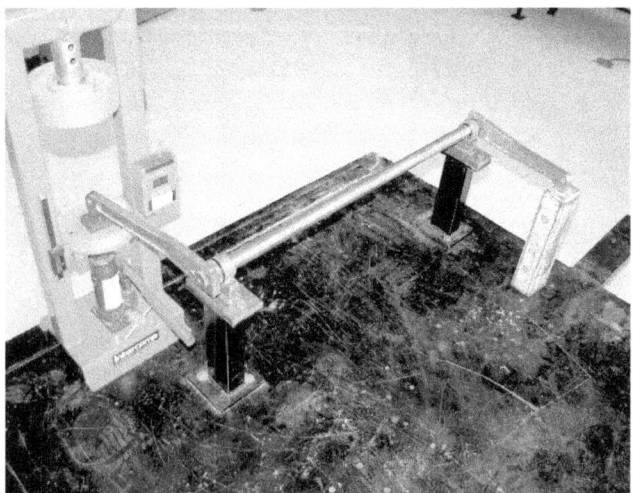

Sway bar systems can be tested for rate. This setup uses a spring rating device to measure pounds of force verses inches of movement of the end of the sway bar arm. It was discovered that the aluminum arm was bending causing loss of rate for the system. A similar steel arm did not bend and provide a higher system rate.

In stock car racing, the sway bar is mounted in front of the spindles and through the use of sway bar arms, it is connected to the lower control arms. As the chassis rolls, the right lower control arm moves up in relation to the sway bar arm and since the end of the sway bar arm is mounted to it, move up too.

In addition to the movement on the right side in a chassis roll, the left side lower control arm wants to move down, but since it is also attached to the other end of the sway bar through its own arm, it cannot easily do that. In this way, the sway bar limits the roll.

The sway bar must twist in order for the chassis to roll. If we use a large enough sway bar, there will be very little roll. It is more effective to place the lower control arm mounts for the sway bar closer to the ball joint where there is more vertical movement.

Eliminating chassis roll completely is not something that we design for. In the recent past, stock car teams have installed very large diameter sway bars that were very stiff in roll resistance. This design did not work very well with the other components of the suspension and setup. Now most teams have returned to using reasonable sized sway bars on stock cars.

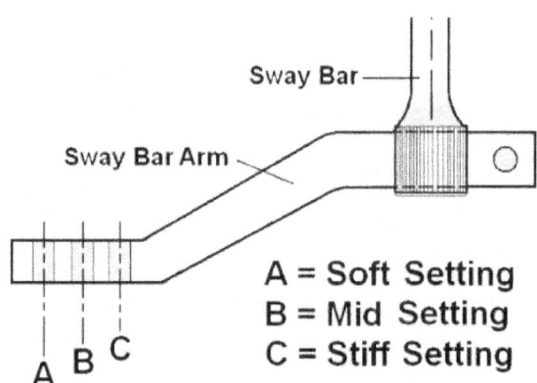

Of the components of the sway bar system that helps determine the rate is the length of the sway bar arm. We can see that the "A" setting provides the longest arm and is the softest setting. As the arm gets shorter by moving the mount towards the "C" hole, the system gets stiffer. The length of the sway bar and the diameter and wall thickness all have an influence on the stiffness of the systems.

Formula Car Anti-roll Mechanisms - For formula and prototype race cars such as the IMSA prototype classes, and Formula One, sway bars are not necessarily bars at all, or may be a combination of a bar and blades. These are linked to the lower control arm movement through a connection to the rocker assembly that transfers the load of the car to the springs.

On this prototype anti-roll system, the links coming from the suspension rockers that move in opposite directions in chassis roll are connected to blades. These blades act much like the arms on stock car sway bar systems. In addition to acting as an arm, the blades will bend and help determine the stiffness of the system. The blades can also be turned. When turned in-line with the links, they will bend very little and offer the stiffest setting. Blades with different thicknesses can be installed to create different roll resistance.

When the chassis rolls, the links on each side move in opposite directions and twist or bend the sway "bar" components. Teams can install different diameter and wall thickness bars and/or different thickness blades to adjust the stiffness of the bar system.

Front Or Rear, Or Both - Circle track stock cars usually don't have sway bars in the rear of the chassis, just in the front. But most road racing and formula cars do have front and rear sway bar systems. That is because the formula cars have rear AA-arm suspensions, a lot similar to the front suspension, that must have roll control.

The stock cars that have rear sway bars are usually racing on road courses where it is hard to control rear roll like they can on oval, circle tracks. The roll stiffness of the sway bars must be coordinated with the spring stiffness.

The adjustment for pre-load, or twisting the bar at ride height, on a late model circle track car is a simple process. This screw jack puts the bar in a bind which loads the bar. Pre-load adds roll resistance before the chassis actually starts to roll.

Exam - In The Context Of This Lesson:

The Difference Between a Sway Bar and Anti-Roll Bar is?

1) One resist sway and the other does not
2) The Sway bar limits wheel movement
3) There is no difference
4) The Anti-roll bar does not roll

A Stock Car Sway Bar Arm Mounts To?

1) The upper control arm
2) The spring mount
3) The chassis frame
4) The lower control arm

Chassis Roll Is Determined Using Which Of The Following?

1) Spring stiffness
2) Amount of lateral force
3) Track width
4) Center of gravity height
5) All of the above

Which Race Cars Usually Don't Use Anti-Roll Bars?

1) Asphalt stock cars
2) Prototype IMSA cars
3) Dirt cars
4) Formula One cars

Which Of These Changes Creates A Stiffer Anti-roll System?

1) Increasing the sway bar diameter
2) Shortening the arm length
3) Increasing the sway bar wall thickness
4) Shortening the sway bar length
5) All of the above

Lesson Thirteen – Double A-arm Geometry

The AA-arm geometry systems in a race car have many components. Here we will describe the parts and how they function. In the Level II Advanced RCT we will get more technical in explaining how each can affect the others. So, let's take a look as the various parts.

- *Chassis Mounts* – These mounts connect the upper and lower control arms to the chassis.

- *Spindle* – The spindle is the connection between the control arms and the wheel. Its function is to allow the wheel to turn and move vertically.

- *Control Arms* – The control arms connect the chassis to the spindle and are free to move with the spindle within the range of motion the car is designed to operate within.

- *Anti-Pro Dive* – Anti and Pro Dive are effects that resist or promote the vertical motion of the chassis when braking.

- *Roll Centers* – The roll center can be two different points. One point is the center of rotation of the chassis and it is referred to as the kinetic roll center. It is formed by the motion of the chassis and has very little to do with race car design.

The other point is the bottom of the moment arm and its location is determined by the control arm angles. This is called the dynamic roll center, or moment center since it represents the bottom of the moment arm for a double A-arm suspension system. The Kinetic roll center is often nowhere near the Dynamic roll center.

- *Jacking Force Theory* – Jacking Force is a theory that states that forces acting through the tire contact patches are the predominant influences in chassis dynamics. This theory has been tested by the author and the results are non-conclusive as to its usability for race car design and setup.

- *Roll Angle Analysis* – Roll angle analysis is a method of comparing the roll stiffness of the front and rear suspension systems on any race car. It is this relationship that controls the load distribution after load transfer occurs during cornering. It is a critical process in determining the actual tire loading values for the four tires.

This is a typical layout of a double A-arm front suspension on a stock car used for circle track racing. We can see the upper and lower control arms are attached to the spindle and the chassis.

Chassis Mounts - The control arms are mounted to the chassis and the spindle. The springs are mounted to the lower control arms. The location of the spring mounts vary with the design. The coil-over (a spring mounted over a shock body to form one piece) chassis mount is above the upper control arm mount and mostly centered between the control arm links for a stock car.

The spring can be mounted as either a coil-over mount, as a large diameter spring housed in a bucket built into the lower control arm, or by way of a push rod that is attached to the lower control arm and extends up to the upper chassis and is connected to a rocker arm that is attached to the spring.

Both the stock big springs and the coil-over springs supporting the front of the chassis are mounted to the lower control arm and therefore the lower control arm must be strong enough to handle the loads and forces it will endure.

The lower control arm chassis mounts can be height adjustable in some race cars. This allows changes to the arm angle from a front view as well as a change to the mounting angle from a side view to assist with anti or pro dive. The upper control arm chassis mount can be adjustable for width, height and side view angle, again for creating anti or pro dive.

The other primary mount at the front of the car is the sway bar mount for a circle track car. These can be loose bushings housed in clamps bolted to the frame for a one-piece type of bar, or a sway bar tube with bearings or bushings at the ends for a three piece bar with the straight bar attached to an arm at each end.

Spindle - The spindle is the connection between the control arms and the wheel. Its design can be matched to the chassis mounting points spacing to accommodate the desired control arm angles. We spoke about spindle design in Lesson Five, but here we will talk about it in relation to geometry.

The primary parts of the spindle are, the ball joints (what the upper and lower control arms are attached to), the spindle snout pin (what the wheels are mounted on), and the steering arms (what the tie rods are attached to).

The spindle pin location from the lower ball joint center of rotation will determine the lower control arm angle aside from the small adjustments some chassis have for the lower chassis mounts. This is called the spindle pin offset. A pin that is placed higher than the normal or customary location is said to be a dropped spindle because it drops the ball joints.

The height of the spindle between the ball joint mounting holes helps determine the upper control arm angles once the spindle pin height has been decided upon. These angles are important in determining the Moment Center location and the camber change characteristics.

Control Arms - In most AA-arm suspension systems, the upper control arm is shorter than the lower control arm. These two arms provide vertical movement for the chassis and resist fore and aft movement of the wheels. The inner mounts are bushings or bearings and the outer connections to the spindle are referred to as ball joints, although the actual mounts can vary from balls to heim joints.

The control arms not only connect the chassis to the spindle, the front view angles they have relative to horizontal determine the moment center height and width. They can be different lengths and offset with the ball joint set rearward from symmetrical to accommodate setting caster.

This lower control arm chassis mount is slotted so that the lower control arm angles can be adjusted. Note that this one is all of the way up. This can help compensate for the high amount of travel associated with the Bump Stop/Spring setups.

Either the upper or lower control arms can be one piece or parts bolted together. If the forward part of the arm is more perpendicular, or at a right angle, to the centerline of the chassis, it could be called a strut instead of an arm.

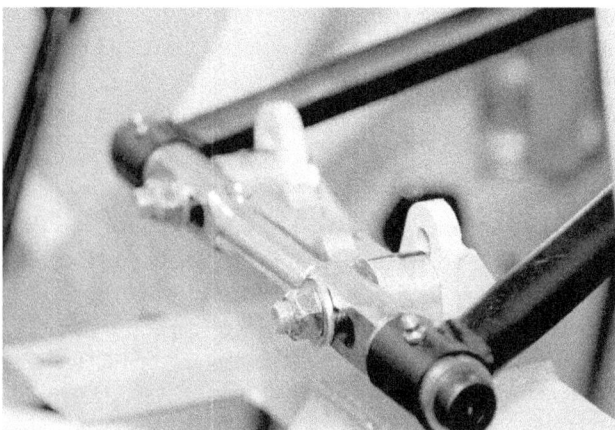

Many race car builders are incorporating slotted mounting holes in the chassis so the racer can tune their arm angles and therefore the MC locations. This Lefthander upper chassis mount has a series of overlapping holes cut in it so that the bolt can be located at different heights to change the upper control arm angles.

The upper control arm on this prototype road race car is very similar to the ones we see on stock cars. The "ball joint" is actually a mono-ball joint. The travel of the spindle on this car is much less than we usually see on stock cars.

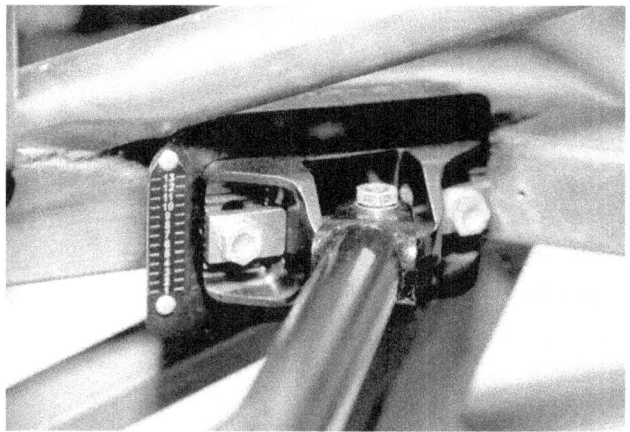

The chassis mount on this prototype race car for the upper mounts in the rear are adjustable for height which will change the control arm angle and the roll center location. The divisions are clearly marked.

Anti and Pro Dive - Anti-Dive is when the forces from braking that are trying to rotate the spindle are used to resist downward motion of the chassis. This effect only exists during braking. In the sketch, we show how when the wheel moves up, as the chassis moves down, with the side view angles in the upper and lower control arms, the ball joints want to move in the opposite direction the force is trying to move them.

It is this contradiction in forces and direction of movement that creates anti-dive. The instant center theory of anti-dive is not relevant to what is actually happening. If it were, then the instant center in most cases is so far to the rear of the front suspension that the lifting force would have little effect.

Pro-dive is when we use these same forces created by braking to assist the suspension in diving. We won't comment on the usefulness of this application, only that it exists. Teams that race on asphalt at circle tracks have been known to use Pro-dive on the left front suspension and anti-dive on the right front suspension.

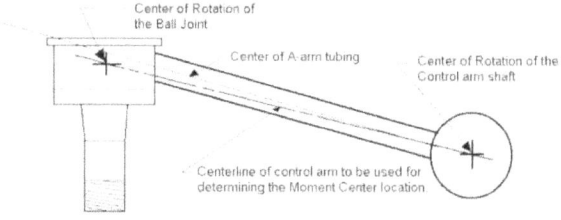

When we refer to the arm angle, we are referring to the angle of a line crossing through the centers of rotation of the ball joint and the chassis mount. The tubing used to construct the control arm may not be inline with a line crossing through the centers of rotation. Therefore, we cannot trust the angle of the tubing used to make the control arm in any roll center calculations.

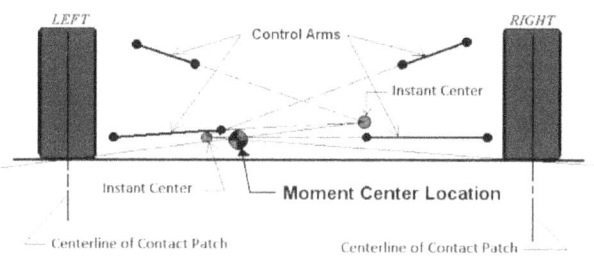

The front Moment Center location is the result using the intersections of lines extended from the centers of rotation of the control arms to form the points we call Instant Centers. The moment center is at the intersection of lines drawn from the left and right IC's to the centers of the contact patches of the corresponding tires.

Roll Center - To be perfectly clear, there are two roll centers engineers talk about for a AA-arm suspension. One is the kinetic roll center that is the point about which a suspension rolls. If you could look at a cross section of the suspension, it would be the point that doesn't move as the chassis dives and rolls going through the turns. That is not the roll center we need to know about.

The other roll center is calculated using the control arm angles, instant centers those angles produce, and intersections with the contact patches of the tire. This roll center is the bottom of the moment arm for that suspension.

Since it is part and parcel with the overturning moment that makes up the dynamic properties of this suspension, we like the term, Moment Center, because it better defines this point and differentiates it from the kinetic roll center. So, from now on in this course, we will refer to this point as the Moment Center, or MC.

The MC is a calculated point in a Double A-arm suspension. It is a point whose location helps to define and control the dynamics of the suspension it is derived from. It is probably the most important part, setup wise, of this type of suspension. Knowing where it is located in both the static (when the car is at rest at ride height) and dynamic (at steady state mid-turn with body dive and roll) locations is a part of the design of a race car just as choosing springs and weight distribution is.

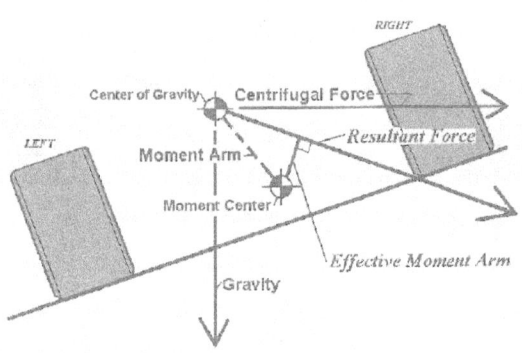

The roll center, or what is better described as a moment center, is the bottom of the moment arm. The top of the moment arm is the center of gravity. The difference in height of the two is the length of the moment arm. This sketch shows how the forces of Gravity and Centrifugal force work in a race car on a banked track.

The MC height represents the bottom of the moment arm in a AA-arm suspension. The top of the moment arm is the Center of Gravity of the sprung mass. More on that later. You have access to a lot of information about roll centers and there are a lot of opinions. I call them opinions because literally none of them have been proven, they just sound right to some people. There is no more controversial part of chassis dynamics than when talking about roll centers.

There has been a lot of study and testing connected with what we just said. In our testing, we have come up with some things we know to be true, and have disproven other theories about how the AA-arm suspension works dynamically.

Here is an explanation of how the front MC location is derived and what affects its location. The sketches are showing both the static and dynamic locations. Moment Center geometry software will help you determine the dynamic location for your MC.

On each side of the car we have upper and lower control arms, in a double A-arm suspension system, and those have pivot points at each end being a ball joint at the spindle and bushings or heim joints at the chassis mounts. If we draw a line through the upper control arm pivot points and again through the lower control arm pivots, these two lines will intersect at a point called an Instant Center.

For most stock cars, these IC intersections lie to the chassis side of the spindle. Each set of control arms on each side of the car has their own IC. If we also draw a line from each IC to the corresponding center of contact patch of the tire on the same side as the control arms we used to create the IC, then the intersection of the two lines from the left and right IC's defines the Moment Center location.

MC Movement – As the chassis dives and rolls through the turns, the control arm angles change, and therefore so does the MC location. We have found, as have many hundreds of race teams and race car builders, that the lateral location of the MC has an important affect on how well the AA-arm suspension performs. Current testing has not determined why exactly this is true, but we think it is due to the camber change associated with the angles that place the MC in a certain location laterally.

Location design for the MC is dependent on which type of race car you are working with. In a road racing car, we have found that if you can keep the MC close to the centerline of the chassis through dive and roll, the car performs much better. Theses cars turn both left and right, so it must work for both turning directions.

If we are working with the front end AA-arm suspension, in a road racing car, we would design the MC to be at centerline statically, and stay close to the centerline dynamically.

For cars that only turn one way, we can design the MC location to migrate farther off the centerline of the chassis and it has been found to work very well, on the race track, this way. So, historically, designers have come up with configurations of control arm angles that will place the MC more to the inside of the turn. When they have done that, the car turns much better than if the MC were more to the outside of the turn.

One many late model dirt cars and fabricated road racing cars, the chassis mount for the upper control arms are uprights of square tubing. To set the height of the chassis mount, holes are drilled in the tubing. To change the upper control arm angle, we only need to drill new holes in the proper location.

Control Arm Angles vs. MC Location - Control arm angles are measured in degrees from horizontal. Therefore, level would be zero degrees and straight up would be 90 degrees. The lower control arm angles largely control the amount of lateral movement of the front MC and the upper control arm angles mostly control the lateral location of the two MC's, static and dynamic.

Excessive lower control arm angles (chassis mounts being lower than the ball joints) cause the front MC to move a greater distance as the car goes through the turns. Increasing or decreasing one or both of the upper control arm angles moves the MC side to side.

Low or reverse (the chassis mount being higher than the ball joint) angles in the upper control arms cause excessive camber change in the front wheels, especially in the right front wheel where it affects the handling the most. In some formula cars, this is common. Because those cars are sprung very stiffly, there is little chassis movement, and therefore less camber change.

In Level II Advanced RCT we will explain how to measure for the MC location and how to redesign a current race car to make changes to the location of the MC.

Exam - In The Context Of This Lesson:

A Coil-over Spring Is Which Of The Following?

1) One that is over the top of the chassis
2) A part of the sway bar package
3) A spring and shock combination
4) Useful for preventing chassis roll

A Chassis Mount Is?

1) The way the engine is connected to the chassis
2) Where the inner ends of the control arms are mounted
3) Anything attached to the chassis
4) 1 and 2

Which Is A Primary Part Of A Spindle?

1) The ball joints
2) Steering arm
3) Snout Pin
4) All of the above

Why Are Control Arms Adjustable For Angle?

1) To assist with anti and pro-dive design
2) To allow changes to the Moment Center location
3) To provide better camber change
4) All of the above

Anti and Pro-dive Only Happen When?

1) The race car is accelerating
2) When the driver is on the brakes
3) When we turn the steering wheel
4) When the roll center is in the correct location

The Moment Center Is?

1) The point around which the chassis rolls
2) Where the control arm angles meet
3) The bottom of the moment arm
4) Used to find shock travels

Lesson Fourteen – Types Of Rear Suspensions

There are two primary rear ends, and accompanying rear suspensions, used in racing. One is the straight axle type that is popular in stock car racing for both circle track and road racing, and the other is the independent rear end used along with a AA-arm suspension that we have already talked a little about.

Each of these have different types of rear end have their unique way that they are connected to the chassis and their own unique methods of spring and shock mounting designs. Let's talk a little about each one starting with the straight axle, so we can become more familiar with the terminology and the various designs.

This rear end unit must be attached to the race car so that it cannot move unrestricted laterally or forward and backwards. This is done through the use of links, or in some cases leafs. The angles these links form can cause movement in all directions of the straight axle when the chassis moves. We'll talk about how that happens later on.

The straight axle has a center section that houses the gears and the differential. This set of gears takes the power from the motor, through the transmission and driveshaft, to the pinion at the front of the differential housing. The pinion is a gear that turns on a ring gear that is attached to the axles through the differential.

The differential is a mechanical device of different designs. It compensates for the different speeds the wheels rotate when we are going around a curve, or through the race track turns. Various designs are named Open differential, Posi-traction, Detroit Locker, Tru-Trac, etc. All of these will help both tires to have Grip and still turn at different speeds from one another.

This is an independent rear suspension. It is independent only because the rear wheels move independent of the gear box. The actual gear case is attached to the chassis and becomes a structural part of the car.

The straight axle rear end is just that, a straight piece consisting of a set of gears, a differential (the part that transmits power from the engine to the wheels), axles mounted to each wheel, axle tubes that house the axles, brake mounts, and control arm mounts. These types of rear end are not a part of the chassis, but rather move independently of the chassis.

There are two different kinds of straight axle rear ends available for stock car racing, the production based stock rear end and the quick change designed primarily for race cars. The most popular stock version rear end is based on the Ford 9 inch design which is considered much stronger. All of the top classes of stock cars touring the U.S. use this 9 inch Ford design.

When a differential is locked with both axles connected together, a spool is used, or the open differential is welded so that the gears are locked. Th spool is a one piece fitting where both axles are contained so that both wheels will turn at the same rotational speed all the time.

The spool would not be of much use on a road course where the rear tires are the same circumference. On oval circle tracks where cars turn the same direction, the spool can be used successfully in some cases because the teams use different size tires for what is called stagger.

In a left turning car, the outside rear tire would be larger in circumference than the left rear tire. The stagger must be matched to the radius and banking angle of the track.

This shows the insides of a quick change type of rear differential. The primary gears are shown, and to the back (left in the photo) are contained a set of easily changed gears that determine the final gear ratio. The primary gear ratio might be something like a 4.11:1 ratio. Then if the quick change gears are say 1.00 to 0.900, then the final ratio would be found by multiplying those two together, for a 4.56 final ratio (1.0 / 0.90 = 1.111 x 4.11 = 4.56). Each gear set has instructions for finding the final ratio using a chart.

The gearing in a rear end can be changed so that for a particular sized track, the engine can operate within a range of RPM (revolutions per minute) that produce the best power. The quick-change rear end allows removal and replacement of the gears without much effort. That is why they are called quick-change. This OEM style rear end must be built to a specific gear ratio and cannot be changed easily. The race team would stock multiple rear ends with ratios they think they might need.

The Straight Axle Rear Suspensions

The various straight axle rear ends use different designs of rear suspensions to locate them within the chassis. The motion of the chassis when entering and driving through the turns will move the links that attach the rear end to the chassis.

This causes changes in the direction the rear wheels are pointed. This is called Rear Steer. Rear steer is a major consideration in the design of a race car, any race car. The Angle of Attack is critical for the rear tires, just as it is for the front tires.

In the rear we have no way for the driver to change the angle of attack like with the front wheels. The only way we can regulate and dictate the amount of rear angle of attack is through design of the angles of the links from the rear end to the chassis.

The most common types of straight axle rear suspensions are:

- ***Three Link*** – This suspension has three links, with two at each side, usually under the axle tube and mounted to the tube and the chassis forward of the tube to restrict fore and aft movement. A third link is attached to brackets mounted above the rear center section forward to the chassis. A panhard bar runs from the rear axle tube laterally to the chassis to restrict lateral movement.

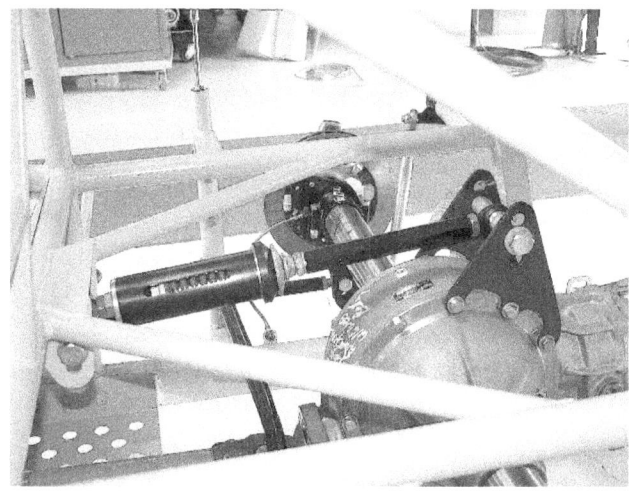

The side view angles the two outer links end up at will largely determine the direction the rear end points when going through the turns. Since the rear tires already have an angle of attack naturally, we just need to fine tune this angle, not radically change it. We will get into this in much more detail in RCT Level Two. The three link suspension is most common on asphalt late model cars.

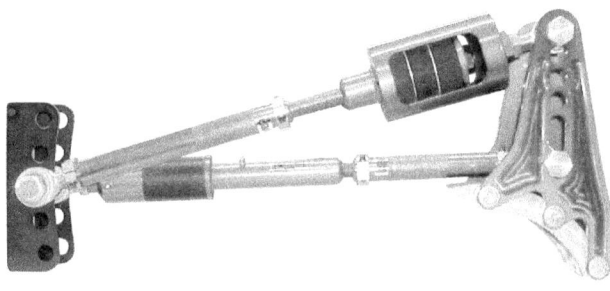

The part of a three link suspension that resists the rotation of the rear end under acceleration and deceleration is the third link which can either be a single link mounted above the rear end, a lift arm mounted to the rear end solidly, or a pull bar that is much like the single link, but which extends using a spring or rubber disc.

- **_Four Link_** – A four link is similar to the three link, except it has two links on each side, one on top of the axle tube and one underneath the axle tube, instead of the just the one on each side for the three link. The four link also has a panhard bar, or what dirt racers call a J-bar, (these two terms are interchangeable) because it is bent into a J shape to go over the drive shaft.

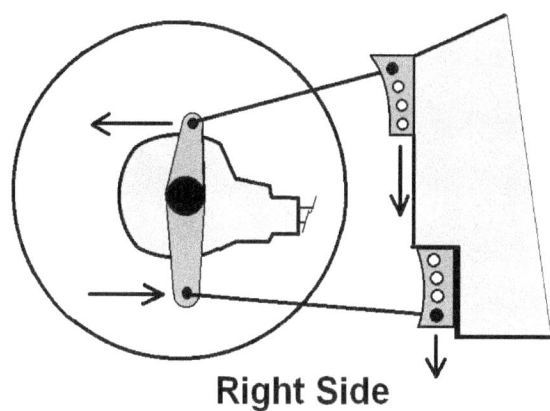

This type of rear suspension is common on dirt cars. Just like the three link, the link angles of the four link are critical in order to get the rear end angle we desire. That ultimate angle is very different for dirt cars than for asphalt cars.

- **_Z-Link_** – A Z-link is a four-link rear suspension where the bottom link runs forward to the chassis and the top link runs back to a mount behind the rear end axle tube. This type of suspension is common on dirt cars, both modifieds and late models. This type also has a third link or lift arm or pull bar to limit the rotation of the rear end. It also has a panhard bar to limit lateral movement.

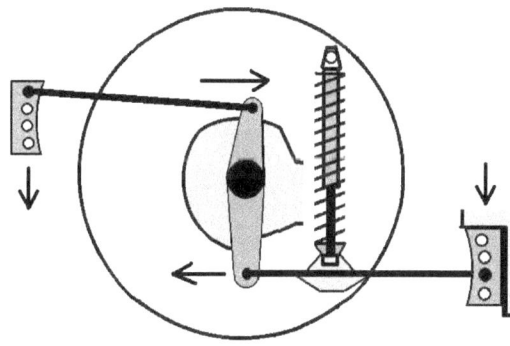

A unique feature of this type is that the spring, or coil-over, is mounted onto the forward link. When this is done, there is a motion ratio created and the chassis for setup purposes, feels less spring rate than the rate of the installed spring.

- **_Truck Arm Link_** – A truck arm rear suspension is based on a mid-1960's Chevy pickup truck suspension. It consists of two arms mounted solid to the rear axle tubes, one on each side, and running forward to bushings that are the chassis mounts. These resist fore and aft movement. This system uses a panhard bar to locate the rear end laterally.

The rear springs are mounted onto the truck arms just forward of the rear axle tubes. Mounting the springs in this way creates a motion ratio much like the Z-link we have already discussed.

- **Leaf Spring** – An older, but still used, type of rear suspension is the leaf spring. This system has two springs, one on each side, that attach solidly to the rear axle tubes, and mount to the chassis using two eyes at each end of each leaf.

The leafs can be single leaf or multi-leaf, meaning a layer of different length leafs are stacked to produce a desired spring rate.

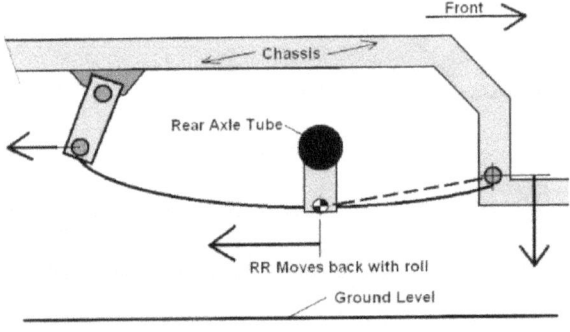

The leaf system provides the springs the car needs in the rear, as well as the lateral locating system for the car in most cases. In some cases, a panhard bar may be used as a lateral locating device.

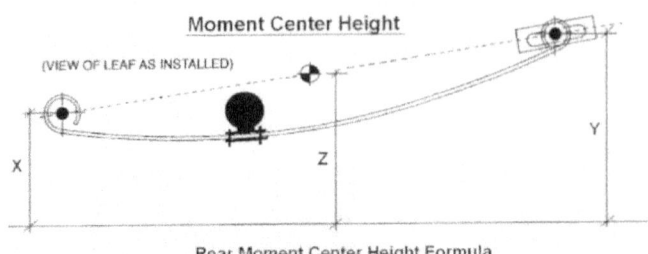

Rear Moment Center Height Formula
$(X + Y) \div 2 = Z$

The leaf syste also provides a means to resist rotational movement of the rear end under acceleration, much like a third link, pull bar or lift arm that are used on a three or four link rear suspension. With the leaf system, there is no need for a third link type of attachment.

- **Metric Four Link** – This suspension type has been used on many 1980's and 1990's production automobiles and is mostly found in stock classes of racing. It consists of four links, two that are mounted on top of the rear end center section, and two at the bottom of the axle tubes that are mounted out farther towards the wheels. Both the upper and lower links run forward to the chassis.

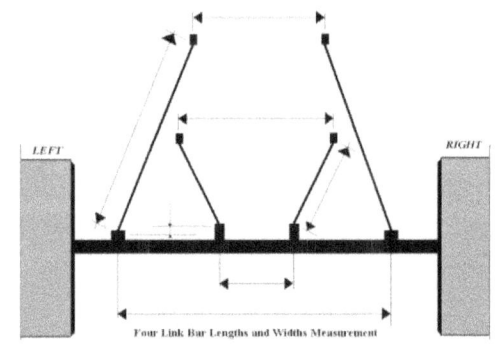

The chassis mounts for the upper links are mounted wider apart than the rear end mounts. And the chassis mounts for the lower links are mounted closer together than the mounts attached to the axle tubes. This matrix of top view angles creates a resistance to lateral movement and this suspension type does not have or need a panhard bar.

Panhard Bar Function

In the straight axle suspensions, the links determine the fore and aft location of the rear wheels. The panhard bar (J-bar used by some classes, but meaning the same) determines the lateral location of the rear end. Because the chassis moves in dive and roll, the angle of the panhard bar is constantly changing.

When the bar angle changes, is moves the rear end slightly one way or the other. We can design movement in to our system to cause the rear wheels to steer, just as we can steer the rear wheels with link angles.

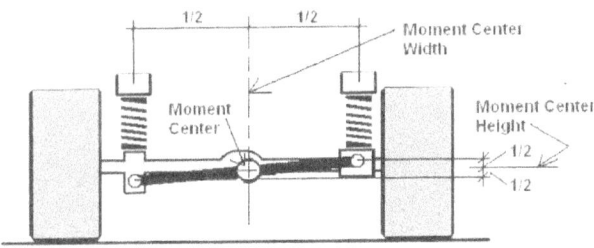

The panhard bar can be mounted to the chassis on either the right side, which is very common for circle track asphalt race cars that use panhard bars. Or, it can be mounted on the left side, which is very common for circle track dirt cars.

For road racing, it can be mounted to either side, but in turning one way, the reaction of the rear end lateral movement will be different when the car steers the other way. So, the panhard bar is not thought to be ideal for road racing applications. This is what is:

Watts Link

The Watts Link is a link that is mounted to both sides of the chassis, one link on top of a center pivot pin, and the other on the bottom of that pin. A plate device is mounted onto the pin and allowed to rotate. When the chassis moves vertically, the plate is free to rotate to compensate for the change in length from the rear pin to each side of the chassis. In this way, the rear end stays centered within the chassis and there is no lateral movement of the rear end that might cause the rear end to steer.

This Watts link was designed and mounted in an early Cup car built by the legendary Smokey Yunick. This car was outlawed at Daytona in 1968 for being way ahead of its time.

Independent Rear Suspension

In most formula and prototype race cars, the rear suspension and rear ends are of the AA-arm type, much like a front AA-arm suspension. The transmission is included within the structure of the rear end and handles the duties of the differential as well. It is commonly called a trans-axle, being a transmission and axle housing. A trans-axle can be ridged in the way the two axles move vertically or independent.

For the independent trans-axle, the most common type, the axles are not connected or housed in a common structure, and can move independently of one another. The difference in rotational speed of the two rear wheels is handled through a differential device that is a part of the transaxle structure.

Just as in a front AA-arm suspension, the rear AA-arm suspension has camber change, a moment center and anti and pro-dive properties. What the rear AA-arm suspension does not need are steering arms or caster, since the rear wheels don't turn.

While the rear wheels don't steer in a rear AA-arm suspension, we do set various toe amounts and there could be bump steer we need to address and minimize in these systems.

Just as much engineering goes into the arrangement of the rear suspension link angles and other geometry parts as for the front suspension. In RCT Level Two, we will explain in more detail how each system works and how to arrange the link angles for maximum performance.

Exam - In The Context Of This Lesson:

The Two Most Common Types of Rear End Used In Racing Are?

1) Lift arm rear end

2) Pull bar rear end

3) Straight axle rear end

4) Independent rear end

5) 3 and 4

The Differential Transfers Power How?

1) From the rear tires

2) Through the third link

3) Through the pull bar

4) From the engine to the wheels

A Straight Axle Rear End Is Used In What Type Of Race Cars?

1) Formula One

2) Dirt late models

3) Asphalt stock cars

4) Stock car touring series cars

5) 2, 3 and 4

An Independent Rear End Is Used In What Type Of Race Cars?

1) Formula One

2) Dirt late models

3) Asphalt stock cars

4) IMSA Prototype cars

5) 1 and 4

Lesson Fifteen – Rear Roll/Moment Center

The rear suspension has a roll center, just as the front does. And like the front, as we have already discussed, we like to call this point a Moment Center because it represents the bottom of the rear moment arm.

The rear MC is determined differently depending on the type of rear suspension. We will look at the three, four, Metric and Z-link systems, as well as the leaf system designed for the straight axle. And we'll examine the rear independent AA-arm system for its MC design.

The Panhard Bar Moment Center

With a suspension that uses a panhard bar as a lateral locating device, the rear MC height is the average height of the centers of the two bolts and/or heim joints that mount the panhard bar to the rear end and the chassis. This height might change when the chassis is moving.

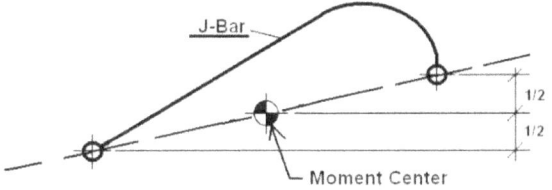

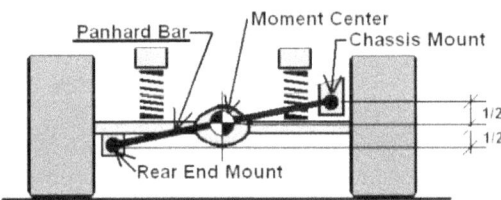

With the Panhard bar or J-bar type of lateral locating device, one end is mounted to the rear axle of the rear end and the other to the chassis. The average height of the two ends is the height of the rear Moment Center, or the height of the front moment center for a car with a straight axle front suspension.

If the panhard bar were mounted to the right side of the chassis, and the car were turning left, then the chassis mount would be lower in the turns due to the motion of the chassis on that side. Since the left side of the chassis moves very little compared to the right, or outside side, the overall MC height will be lower in the turns.

On this rear axle tube mount, there is a serrated plate with a matching through bolt that makes the height of the rear end mount adjustable for height. In this way the team can fine tune the rear moment center height. There is usually a similar adjustable mount on the chassis side.

Remember that the rear MC for a panhard bar system is the average height of the two ends. If one end is lower in the turns, then the average height of the MC will then also be lower.

If the panhard bar were mounted to the left side of the chassis, in a left hand turn, then both sides would remain at nearly the same height and the height of the MC would not change.

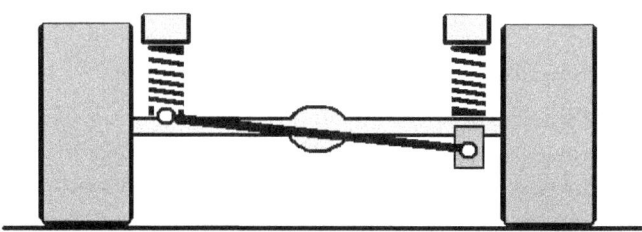

The current trend in circle track racing where the cars turn left is to mount the right side of the panhard bar lower than the left side axle mount. This puts the bar more inline with the combined forces of Gravity and lateral force through the turns. It provides a more straight pull on the bar which reduces possible adverse jacking forces.

For road racing applications, in a left turn, with a left chassis mount, the MC height would stay the same, but turning the other way, it would change and be lower. This causes different handling balance from left turns to right turns, which is not ideal.

The Leaf Spring Moment Center

With the leaf spring system, the rear roll center is calculated in the following way. We average the height of the two mounting eye bolts at the front and rear of each spring. Then we average the two right and left average heights of the eye bolts and that is the height of the rear MC for leaf springs.

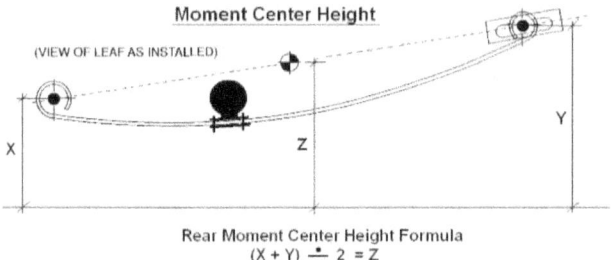

This is the formula for calculating the moment center height of a straight axle, leaf spring suspension. The rear moment center height is the average of the heights of the two springs individual moment center height, in case they are not the same. As the chassis rolls and the outside chassis moves lower, that side will produce a lower point in the moment center package which will lower the overall moment center height, much like a panhard bar does when its chassis mount moves lower on the right side.

With the leaf system, we also have a change in MC height when we are going through the turns much like a panhard bar, except that we have the same change in height when turning both ways. The car stays consistent between left and right hand turns.

To change the height of the rear MC for a leaf system, we need to change the height of the spring in relation to the ground, and we do that by spacing the spring differently at the mount to the rear axle tube.

If we want the rear MC to be higher, but the chassis to remain the same height, then we need to make two changes. First we raise the leaf spring in relation to the rear end axle tubes. Then we raise the spring mounts on the chassis by the same amount. We have then raised the MC height without raising the height of the chassis.

The Watts Link Rear Moment Center

The Watts link rear suspensions Moment Center is unique in that it stays very consistent in height. This is because the link is mounted to the rear end and stays relatively at the same height throughout the lap. We change the moment center height by changing the height of the swivel link, and ideally the chassis mounts of the links from the centerpiece to the chassis so that they remain level.

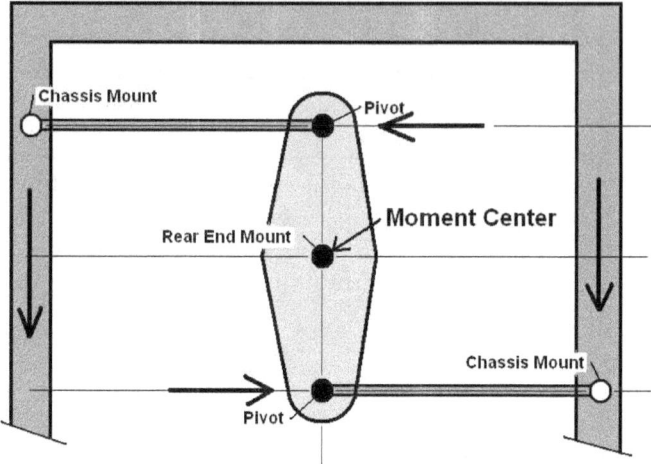

The Watts link has a moment center located at the center of the pivot bracket. This point can, in some designs, be moved up or down to tune the height of the rear moment center.

Most designs of Watts link have incorporated these adjustments into the overall design of the rear suspension. Otherwise, there is no way to change the moment arm in the rear to tune the balance between the front and rear suspensions.

The Independent AA-arm Rear Moment Center

The MC for an independent rear suspension using a AA-arm geometry is the same as when we are measuring a front end AA-arm suspension. We use the upper and lower arm angles to produce an instant center for each side. Then the lines from the instant centers to the contact patch on each side are crossed. The point where they cross is the rear MC location.

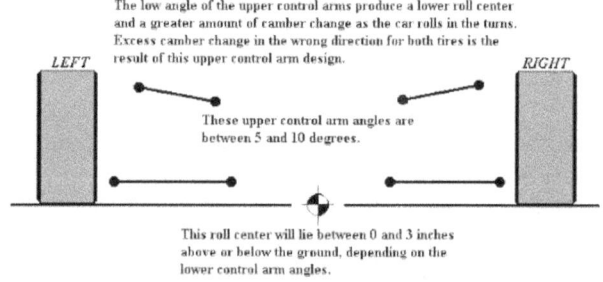

In a Double A-arm rear suspension, there are control arms much like the AA-arm front suspension. The moment center is adjustable in most of these systems by changing the angles of the control arms at the chassis mount.

For design purposes, we want the rear MC to be centered at static ride height and to not move very far from the centerline when the car is going through the turns. We can do this by arranging the arm angles so that after plotting or computing the MC location after dive and roll, the MC stays near the center.

Exam - In The Context Of This Lesson:

The Rear Moment Center Is?

1) The point around which the rear of the chassis rolls

2) The top of the rear moment arm

3) The bottom of the rear moment arm

4) The center of the rear chassis at static ride height

The Panhard Bar Moment Center Is?

1) Likely to move in the turns

2) Very stable in height when the chassis moves

3) The average height of the end mounting bolts

4) 1 and 3

The Leaf Spring Moment Center Height Is?

1) Where the spring mounts to the rear axle tube

2) Averages of the center of the eye mounts for the two springs

3) The front end heights of the leaf mounts

4) The center of the axle tubes

The Independent AA-arm Rear Moment Center Is?

1) The average of the instant center heights

2) Always located at the center of the chassis

3) Calculated by the same method as the front AA-arm MC

4) Should always be to the outside of the turns

Lesson Sixteen – Alignment and Toe

Alignment for the rear suspension is looked at a little differently than the way we looked at the front. This has a lot to do with what we talked about in Lesson Two, Why A Race Car Turns. If you remember, our rear wheels are already pointed to the left of the tangent line we are following around an arc, or turn of a race track.

Rear Steer - First off, the rear wheels do not turn, or at least the driver cannot steer the rear wheels. The links that attach the rear end to the chassis can cause the rear wheels to steer when the chassis moves. This movement is called Rear Steer.

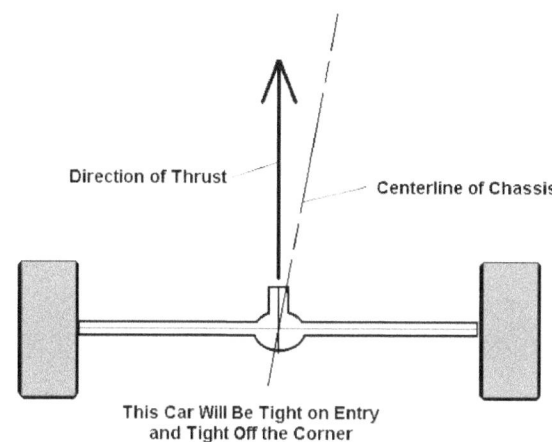

This Car Will Be Tight on Entry and Tight Off the Corner

For most applications, a small amount of rear steer to the inside of the turn can help gain grip. If we overdue the steer in this direction, we can make the car tight. The example is related to left turning circle track racing.

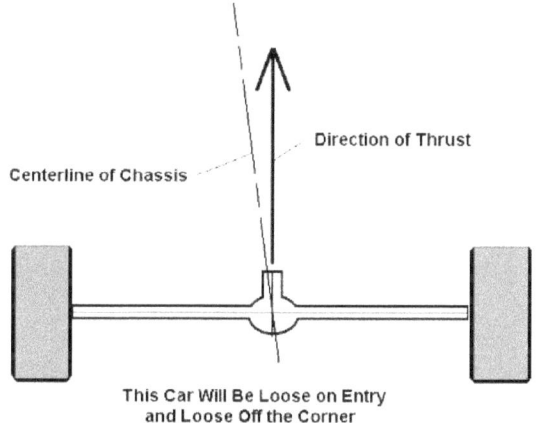

This Car Will Be Loose on Entry and Loose Off the Corner

Rear steer can be detrimental to performance. We can use some amount of rear steer to our advantage, but excess steer is never an advantage. Steer to the right will make a circle track car loose, or in the case of a road racing car, loose in the left hand turns and tight in the right hand turns. This miss-alignment can come from static settings or from movement of the suspension parts causing the steering.

The steering direction can be either to the outside, or left, of the tangent line, or to the right of it. When the steering is to the right, we call this Rear Steer to the right and the opposite when the wheels are steered to the left of the tangent line.

For the effect of RS, we need to think about which way the car is turning. In a left turn, such as in circle track racing in the U.S., RS to the right would make the car loose, or oversteer depending on the amount of the change in the steering angle.

For left hand turns, RS to the left would tighten the car, or make it understeer is the amount of steering angle change were more than the amount the rear wheels needed to resist the lateral force, or centrifugal force, which tries to push the rear of the car to the outside of the turn.

For turning only one way, we can easily design RS into our race car that will maximize and optimize our rear wheel/tire angle of attack to work in conjunction with the front wheels/tires. This helps to balance the setup not only for handling balance, but for dynamic balance too, our primary goal for race car setup.

In RCT Level Two, we will get into a lot of detail about how different systems produce RS and how we can design and tune the different suspensions. We will help the student understand the limits to RS and how the designs are different for different classes of race cars.

For now, just know that there is a fine line between too much and too little RS. We want to develop a good amount of RS to produce the highest tire angle of attack without going too far and losing all of our rear grip.

Rear Alignment – The rear alignment for race cars is fairly simple, but there are variations that teams choose to follow. Simply put, for circle track racing where we are only turning one way, the outside tire contact patches should follow the same line.

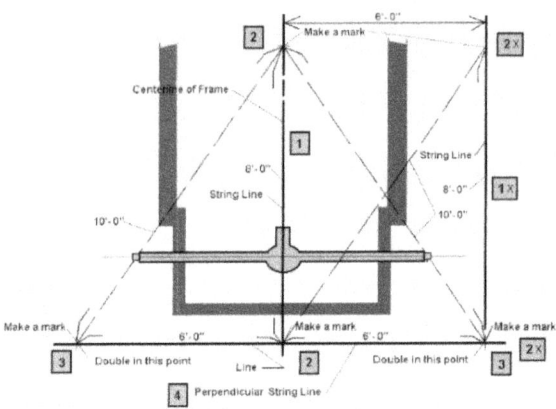

To setup for alignment of the rear end, we need to establish alignment of the chassis and find the centerline or some datum line to use as a reference. Here, in the example we use a perimeter car that is symmetrical. Once the centerline has been established, we setup a line that is ninety degrees off of the centerline to check the rear end for square.

If the front and rear track width (distance between the centers of each wheel contact patch on one end of the car) are different, then we line up the outside tire contact patches and the inside contact patches will be miss-aligned.

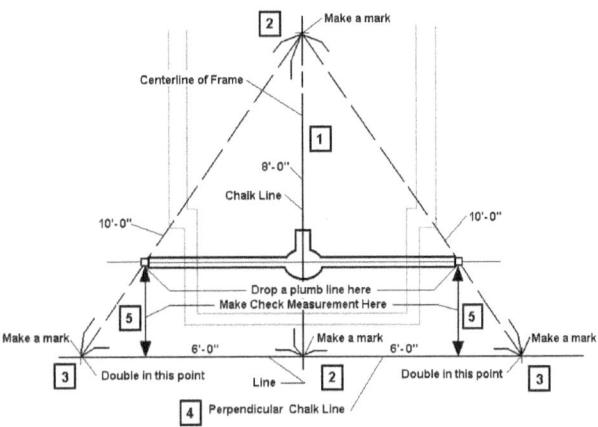

With the rear line established off of the centerline, we measure to the outer ends of the rear end to see if it is square. Adjustments to the lengths of the suspension links will bring the rear end into square.

We do that because the outside tires will carry the most loading in almost every case. The more heavily loaded tires will perform better when they are inline and running on the same arc.

For road racing cars, we center the front and rear tire contact patches on the chassis centerline. If the front and rear track widths are different, then the outside and inside tires will be miss-aligned equally. We also make sure that all four wheels are pointed parallel to each other. All tires would be parallel to the centerline of the chassis.

Rear Toe – Rear toe for a straight axle rear suspensions, running on circle tracks, is zero in most cases. We want to toe these wheels straight ahead so the tire angle of attack is consistent for both tires.

For road racing, we often see where teams will toe the rear tires in (the front of the tire is narrower than the rear of the tire). This is opposite of the way we mostly toe the front tires, being toe-out. This is done for stability mostly. On high speed long straights, the rear of a road racing car can dance around somewhat.

Rear toe can be an issue as well. If the rear wheels are not parallel, then we can have toe-in or toe-out in the rear wheels for a straight axle suspension. This rear end has a brace installed that can help it keep in alignment and correct any miss-alignment that may occur.

This design goals for toe-in in the rear for road racing cars is easy to do with a AA-arm rear suspensions. There is a tie rod link that positions the rear spindles so that the rear wheels are pointed where we want them to be pointed. We can easily make changes to the tie rod lengths to make changes to the rear wheel toe settings. Normal toe amounts for the rear are from zero toe to 0.050" of toe per wheel. We usually don't see much more that that used.

To create toe for a straight axle suspension, we need to bend the axle tubes. We can do this by heating them on one side, an art form at least, or by installing toe links that will pull on the front of the axle tubes to produce toe-in. These links will keep the toe consistent as the car is raced and can be easily adjusted.

Exam - In The Context Of This Lesson:

What Mostly Miss-aligns The Rear Of The Race Car?

1) The driver using his foot on the throttle
2) Incorrect lengths of the rear suspension links
3) Rotation of the rear end
4) Squatting of the rear suspension

Rear Alignment To The Left Would Do What?

1) Cause the car to be slower
2) Make the car faster
3) Point the tires left in relation to the front tires
4) Make the car looser

Rear Alignment To The Right Would Do What?

1) Cause the car to be slower
2) Make the car faster
3) Point the tires right in relation to the front tires
4) Make the car looser

Rear Alignment Can Do What?

1) Help the rear to gain grip
2) Cause the rear to lose grip
3) Help the rear track more inline with the front tires
4) All of the above

For Road Racing, We Align The Rear Tires How?

1) Line up the outside tire contact patches
2) Line up the inside tire contract patches
3) Miss-align both rear tires the same
4) Make both rear tires line up with the front tires

Rear Toe Is Usually?

1) With the tires pointed out at the front
2) With the tires pointed in at the front
3) Straight ahead
4) 2 and 3

Lesson Seventeen – Rear Steer, Roll Steer and Bump Steer

Alignment for the rear suspension is looked at a little differently than the way we looked at the front. This has a lot to do with what we talked about in Lesson Two, Why A Race Car Turns. If you remember, our rear wheels are already pointed to the left of the tangent line we are following around an arc, or turn, of a race track. So, we don't really need to steer the rear wheels unless we think we can gain performance by doing so.

The most extreme case of rear steer can be seen in a dirt late model or dirt modified. There are advantages that can be realized on dirt where these same tactics would never work on asphalt race tracks. Here we will discuss the conditions of rear steer and roll steer and how they occur.

Rear Steer - First off, the rear wheels do not turn, or at least the driver cannot steer the rear wheels. The links that attach the rear end to the chassis can cause the rear wheels to steer when the chassis moves. This movement is called Rear Steer.

We talk about Rear Steer mostly in relation to circle track cars. For road racing cars, there may be some slight amount of rear steer incorporated into the designs of the rear suspension, but nothing compared to the amount used for some circle track cars, mostly dirt cars.

This is a dirt car rear suspension left rear link system. There are numerous holes in the chassis to allow the team to change the angles of the links. Depending on the angles, when the chassis moves in bump or roll, the rear end on that side of the car will move forward, backward, or not at all. These types of suspension add a high degree of complexity to the setup of dirt cars not seen in any other type of race car.

The steering direction can be either to the outside, or left, of the tangent line, or to the right of it. When the steering is to the right, we call this Rear Steer to the right and the opposite when the wheels are steered to the left of the tangent line.

For the effect of RS, we need to think about which way the car is turning. In a left turn, such as in circle track racing in the U.S., RS to the right would make the car loose, or oversteer depending on the amount of the change in the steering angle.

This right rear chassis mount for the three link rear suspension on an asphalt late model car is adjustable for height. This adjustment changes the angle of the link. When the chassis rolls, the angle of the link can either pull the right rear wheel forward, backwards or have it remain where it was after the car has fully rolled. During the roll process, this link will, in most cases, push the wheel back before it pulls it back towards the front.

For left hand turns, RS to the left would tighten the car, or make it understeer is the amount of steering angle change were more than the amount the rear wheels needed to resist the lateral force, or centrifugal force, which tries to push the rear of the car to the outside of the turn.

For turning only one way, we can easily design RS into our race car that will maximize and optimize our rear wheel/tire angle of attack to work in conjunction with the front wheels/tires. This helps to balance the setup not only for handling balance, but for dynamic balance too, our primary goal for race car setup.

In RCT Level Two, we will get into a lot of detail about how different systems produce RS and how we can design and tune the different suspensions. We will help the student understand the limits to RS and how the designs are different for different classes of race cars.

For now, just know that there is a fine line between too much and too little RS. We want to develop a good amount of RS to produce the highest tire angle of attack without going too far and losing all of our rear grip.

Roll Steer – Rear roll steer is much the same as bump steer. In a rear suspension, the links that hold a straight axle in place, or the control arms in a AA-arm, might steer the rear wheels when the car rolls going through a turn. We can eliminate this roll steer, or we can use it to our advantage to gain grip or overcome other handling problems.

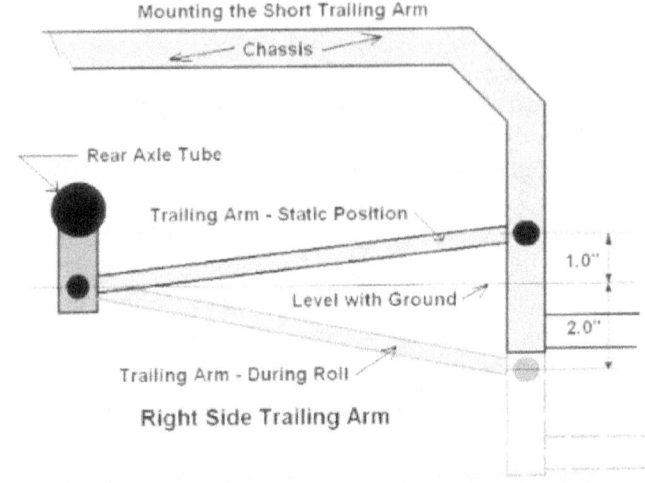

A common way to set the right rear link, or trailing arm as it is called, chassis mount is to set it up from level by one third of the total distance it will travel due to chassis roll. This will provide a slight amount of rear steer to the left for more angle of attack for the rear tires and more grip.

Depending on how the links are setup, we can cause rear steer to develop in any direction, right or left of centerline. From what we have already learned in Lesson 16, steer to the right would loosen the car and steer to the left (in a circle track car turning only left) would tighten the car.

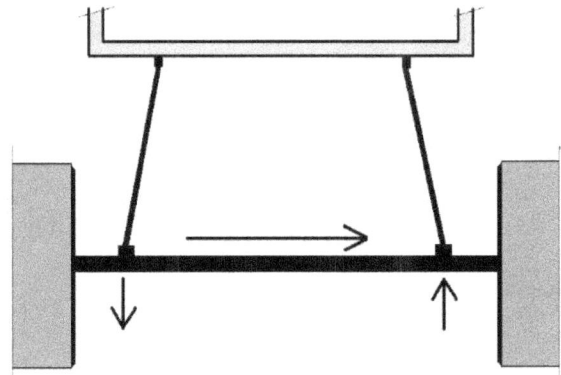

On a race car with a panhard bar, the bar does not remain level during chassis roll. Because it is attached to the chassis, and the chassis moves vertically, it will pull the rear end one way or the other. On some designs of rear suspension, the links are not parallel from a top view. So, when the rear end will move side to side as the panhard bar moves through its travel, creating rear steer.

Bump Steer – Bump steer in the rear suspension is much the same as in the front AA-arm suspension. When the chassis moves vertically, the wheels steer if there is any inherent bump steer. We try to eliminate bump steer in the front suspension, but in the rear, we might want bump steer in some cases.

If we desire a straight ahead alignment when going through the turns then the roll steer characteristics must be such that no steer occurs. But if we want rear steer coming off the turns, then we can introduce bump steer to steer the rear end when the rear squats under acceleration. This is accomplished with the inside links on a circle track car. We'll get more into detail on how that is done in RCT Level Two.

There are many ways to introduce rear steer in a race car. One way that some teams use is to include a link in the right rear that is allowed to compress under acceleration. This compression shortens the link and pulls the right rear wheel forward to introduce rear steer to the left of centerline to tighten the car.

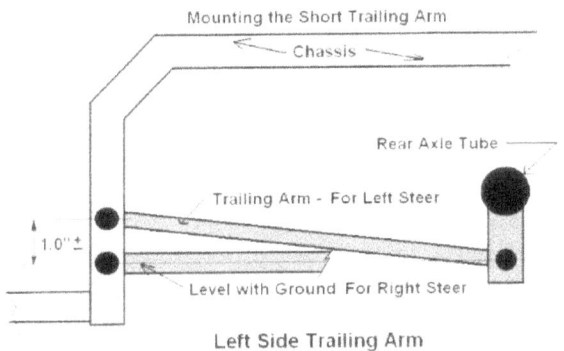

We get bump steer in the rear suspension from the car squatting upon acceleration off the corners. The rear of the car will squat and travel down some amount depending on other design features, and when it does, we can produce rear steer if the links are arranged in a certain way. If the left rear link in a three link suspension is set with the chassis mount higher than the axle mount, when the car squats, it will push the left rear wheel back. This creates rear steer to the left of centerline.

Rear Alignment – The rear alignment for race cars is fairly simple, but there are variations that teams choose to follow. Simply put, for circle track racing where we are only turning one way, the outside tire contact patches should follow the same line.

If the front and rear track width (distance between the centers of each wheel contact patch on one end of the car) are different, then we line up the outside tire contact patches and the inside contact patches will be miss-aligned.

We do that because the outside tires will carry the most loading in almost every case. The more heavily loaded tires will perform better when they are inline and running on the same arc.

For road racing cars, we center the front and rear tire contact patches on the chassis centerline. If the front and rear track widths are different, then the outside and inside tires will be miss-aligned equally. We also make sure that all four wheels are pointed parallel to each other. All tires would be parallel to the centerline of the chassis.

Rear Toe – Rear toe for a straight axle rear suspensions, running on circle tracks, is zero in most cases. We want to toe these wheels straight ahead so the tire angle of attack is consistent for both tires.

For road racing, we often see where teams will toe the rear tires in (the front of the tire is narrower than the rear of the tire). This is opposite of the way we mostly toe the front tires, being toe-out. This is done for stability mostly. On high speed long straights, the rear of a road racing car can dance around somewhat.

This design goals for toe-in in the rear for road racing cars is easy to do with a AA-arm rear suspensions. There is a tie rod link that positions the rear spindles so that the rear wheels are pointed where we want them to be pointed. We can easily make changes to the tie rod lengths to make changes to the rear wheel toe settings. Normal toe amounts for the rear are from zero toe to 0.050" of toe per wheel. We usually don't see much more that that used.

To create toe for a straight axle suspension, we need to bend the axle tubes. We can do this by heating them on one side, an art form at least, or by installing toe links that will pull on the front of the axle tubes to produce toe-in. These links will keep the toe consistent as the car is raced and can be easily adjusted.

Exam - In The Context Of This Lesson:

What Mostly Steer The Rear Of The Race Car?
1) The driver using his foot on the throttle
2) Incorrect Angles of the rear suspension links
3) Rotation of the rear end
4) Squatting of the rear suspension

Rear Alignment To The Left Would Do What?
1) Cause the car to be slower
2) Make the car faster
3) Point the tires left in relation to the arc tangent line
4) Make the car looser

Rear Alignment To The Right Would Do What?
1) Cause the car to be slower
2) Make the car faster
3) Point the tires right in relation to the arc tangent line
4) Make the car looser

Rear Steer Can Do What?
1) Help the rear to gain grip
2) Cause the rear to lose grip
3) Help match the front grip levels for more balance
4) All of the above

For Road Racing, We Align The Rear Tires How?
1) Line up the outside tire contact patches
2) Line up the inside tire contract patches
3) Miss-align both rear tires the same
4) Make both rear tires line up with the front tires

Rear Toe Is Usually Set This Way?
1) With the tires pointed out at the front
2) With the tires pointed in at the front
3) Straight ahead
4) 2 and 3

Lesson Eighteen – Anti-Squat

Anti-Squat is just what the term implies. It is an effect caused by the force that is accelerating the race car, and that force uses the links that position the rear end to provide Anti-Squat (A/S). There has been a lot written about anti-squat and we are going to try to explain how it really works. This seems complicated, but in reality, it is a simple concept. Refer to the drawings often to try to picture what is going on here.

When the engine is powered up to full throttle (or something in-between), the torque goes through the transmission, drive shaft and rear end gears to turn the axles and rear tires to propel the car forward and to make it speed up.

In every rear end that is based on a straight axle design and where the transmission, or transaxle is mounted to the chassis or engine, this effect is a part of the geometry. So, we are talking about circle track cars as well as road racing cars.

Anti-squat is a mechanical effect in the rear suspension of a race car that initiates a force that counters the tendency to squat, or compress the rear springs, upon acceleration. As the car accelerates, weight is transferred to the rear suspension and that causes the rear spring to compress. Anti-squat reduces that motion.

Straight Axle Rear Ends - In those cars with straight axle rear ends, the drive shaft connects to a pinion shaft and pinion gear that turns a ring gear. The ring gear is connected to the axles and those are connected to the wheels. It is the action of the pinion gear turning the ring gear that creates part of the force needed for AS.

The rest of the force is provided by the thrust, or pushing forward as the car accelerates.

To propel the car forward, while the pinion gear is turning the ring gear, it is trying to climb the ring gear. In doing so, it is also putting a Torque force on the rear end trying to rotate it so that the pinion shaft would move up, if it could. There is a lot of force involved with this action.

Note that we said this only happens when we are applying power from the engine through the drive shaft and rear end. As the pinion tries to climb the ring gear, it wants to move up causing a powerful force that is also trying to lift up on the pinion and rotate the rear end in a clockwise direction when viewed from the left side. Try to imagine this force for a moment.

At the same time, the Thrust force is pushing on the suspension links. It is the combination of those two forces, engine torque to the rear end in a straight axle car, and forward thrust in all types of rear ends. that cause A/S to occur.

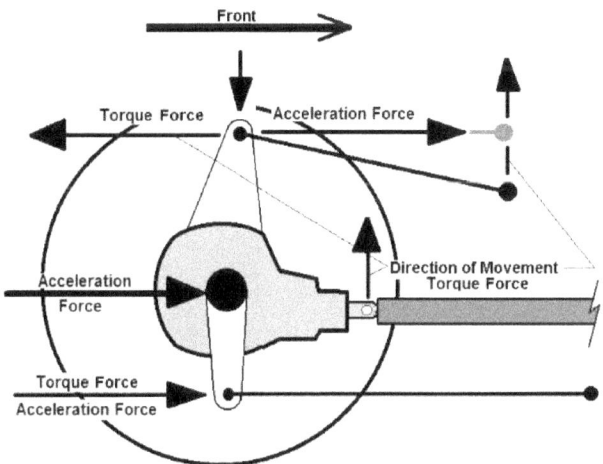

This shows how anti-squat works in three link rear solid axle suspension. As the car accelerates, the torque of the engine is transferred to the rear end and tries to rotate the rear end. As this rotational force is applied, the bottom links are pushed into compression and the top third link is put in tension, or pulled against. At the same time, the Acceleration forces are pushing against both the upper link(s) and the lower links. At the top, the forces are opposed and at the bottom they combine.

If the links we attach to the rear end to keep it from rotating are arranged in a certain way, this force will try to lift the rear of the car. The force trying to lift the car has to overcome the weight of the rear portion of the sprung weight as well as one other load.

The other load is this. When we accelerate the car, we create load transfer from the front to the rear. This load is easily calculated, although we won't be doing the calculations here. We just want you to understand the process and what is involved.

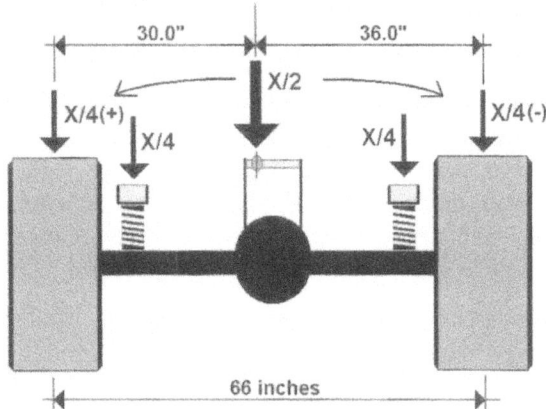

As weight is transferred to the rear upon acceleration, the anti-squat effect lifts on the front of the three links in a three-link suspension and pushes down on the rear end at the point where the three links are mounted to the rear end bracket. If we move the third link mount left, the, its load is applied more to the left and closer to the left rear tire. The same applies to the lower links (not shown) in that they cause a lifting force to help overcome chassis squat upon acceleration and put an equal loading onto the rear axle housing.

So, we have load transfer to the rear when we accelerate and that added load would obviously cause the rear of the car to be lower since the springs will have to compress when that added load is put upon them. With A/S, we can prevent the springs from compressing by taking the added load onto the links through A/S geometry.

Third Link - The link that helps cause AS can be a "third link", a lift arm, a lift bar, or other types of link that prevents the rear end from rotating. In the case of the third link, it is mounted with the front end lower than the rear end. The front is mounted to the chassis and the rear connection is mounted to the rear end housing.

As the rear end tries to rotate from acceleration, it pulls on the third link. Some of these links that use springs to allow them to expand are actually called pull bars. As the force is put on the third link, it wants to become straight to the direction of the force. In doing so, it creates a lifting force at the chassis end that makes it want to move up.

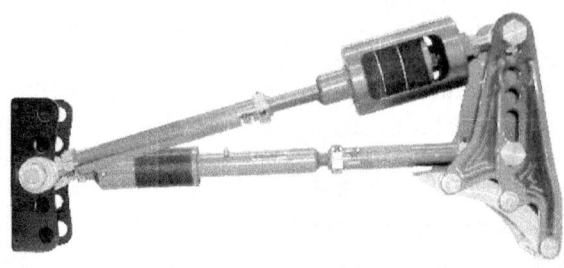

Various designs of third links are available and do different things. This one applies the anti-squat forces, reduces the torque shock to the rear end, and reduces adverse wheel hop when the car is braking. Note the angle of the pull-bar on the top. This angle is what creates the anti-squat force. The bottom link is used for compression resistance when decelerating.

The front end of the third link mounted is lower and wants to move up. This is a part of the lifting action that reduces rear squat that is the A/S component in a three link rear suspension. The rear of the third link wants to go down in order to become straight to the pull and this puts loading on the rear end that is centered where the link is mounted.

So, A/S not only helps prevent squat, it also puts some of the transferred load onto the rear end and not onto the springs where it would go if there were no A/S. We can use this A/S effect to re-distribute loading on the rear tires while the car is accelerating. By moving the third link sideways, and/or changing the angles of all of the links, we can change the load distribution on the rear tires. We will get more into that in RCT Level Two.

Lift Arm/Bar - With the Lift arm type of third link, we have a different application of the force referred to the rear end. We still have the torque through the driveshaft and pinion, but when the rear end tries to rotate, it is resisted by an arm mounted solid to the rear end housing that tries to lift the sprung weight while putting additional loading onto the rear end at the same time.

This type of link is more adjustable for the amount of force that represents A/S by being adjustable for length. A longer arm or bar creates less A/S force. This is pure leverage at work. The two different types of link will provide A/S in similar ways, but it is the adjustability that differentiates them.

The lift arm (black tubing beside the drive shaft) is another design of anti-squat used mostly in dirt cars. As the car accelerates, the arm, mounted solidly to the rear end, tries to lift the chassis at the point where the spring and shock are mounted. The opposite direction of force is applied to the rear end again at the point where the lift arm is mounted.

The Lower Links – While all of that is happening with the upper links in a three link system, the lower links, if the angles are pointing up going forward, will themselves provide A/S forces. The Torque of the engine transferred to the rear end will provide a force onto all of the three links. The Thrust forces also act on the three links, but for the lower links in a three link suspension, the Torque and Thrust forces combine and this makes the A/S forces and effect more significant for the lower mounts.

In the case of a lift arm car, all of the Thrust is taken by the four links in this suspension system. Since the lift arm controls all of the torque forces, then all of the Thrust forces are taken by the four links. It is the arrangement of the link, or bar, angles that determines the A/S effect for those cars.

It is possible, and quite common, for four-link cars to create more than 100% A/S and lift the rear of the car higher than it would be at ride height. Many dirt late model setups lift the left rear corner higher than the right rear corner due to the angles of the left side bars. This puts a lot of loading, sometimes most of the rear loading, onto the left rear tire.

Exam - In The Context Of This Lesson:

Anti-squat Is Caused By?

1) Braking into the corners
2) The torque of the motor when accelerating
3) Forward thrust due to acceleration
4) Stiff rear springs
5) 2 and 3

Anti-squat Effect Is Common In?

1) Prototype race cars with transaxles
2) The front suspension of a race car
3) Straight axle rear end suspensions
4) Cars with high centers of gravity
5) 1 and 3

Anti-squat Produces What Effect?

1) It tries to lift the rear of the car
2) It causes an effect that adds load to the rear tires
3) It helps the car to turn the corner
4) It creates a force that resists the effects of load transfer
5) 1 and 4

The Length Of A Lift Arm Affects What?

1) The amount of load transfer under acceleration
2) Acceleration off the turns
3) The amount of anti-squat force
4) Torque transferred to the rear end

Lesson Nineteen – Driveline Alignment

Over View - A Driveline is a connection between the engine and the rear end in a race car with a straight axle rear suspension. Because the rear end moves in relation to the chassis, and because the engine is mounted solid to the chassis, there must be a connection that can adjust to the motion between the motor and the rear end.

Drive Line Terminology – The component names for the driveline parts are as follows:

Slip Yoke – this is the yoke at the front of the driveshaft that goes into the transmission and can slip to allow adjustment for chassis movement.

Weld Yoke – is the yoke at the each end of the drive shaft that accepts the U-joint. These attach to the both the pinion yoke at the rear end and the Slip yoke at the front. They are bonded to the shaft tubing.

Universal Joint Kit or U-Joint (UJ) – is the actual part that forms the rotational connection between the driveshaft and both the transmission and rear end.

Tubing – is the metal or composite tubing between the yokes.

Pinion Yoke – is the yoke that is attached to the pinion shaft at the rear end.

How A Drive Line Should Work - A drive shaft has two Universal joints, called that because they can move universally, in any direction up to a certain degree of angle. The front UJ is connected to a yoke that slides over the transmission output shaft. The rear UJ connects the drive shaft to the pinion shaft at the rear end.

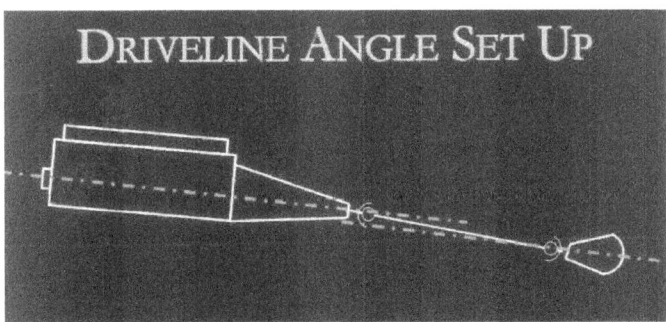

Driveline alignment is a very simple concept. It has been firmly established long ago what the correct alignment should be in order to reduce driveline vibrations and parts fatigue. It is, the engine/transmission must be inline and parallel to the pinion shaft in the rear end. And, the angles should be minimal between the engine/transmission/pinion shaft and the drive shaft.

The angles the transmission output shaft and the pinion shaft make with the driveshaft is critical. One of the most important design parts of a high performance race car is drive line alignment. And it is very simple in concept.

Ideal Drive line alignment features are: 1) Having the transmission output shaft parallel to the pinion shaft; 2) Making sure the angle created between both the transmission output shaft and the pinion shaft to the driveshaft is minimal. It is that simple, but again, very important.

Drive Line Angles Gone Wrong – When the UJ operates with any amount of drive line miss-alignment, it creates a problem. The bearings speed up and slow down twice per revolution of the driveshaft. This causes an oscillation in the power train. The more miss-alignment angle that we have, the higher the peaks of oscillation we see and therefore the greater chance of vibration.

Drive line miss-alignment angles are a cause of vibration and power loss. If your race car must have driveline angles from a design standpoint, the angle of the drive shaft to both the transmission output shaft and the pinion shaft should be equal and also opposite. The angle should also be kept to a minimum when at all possible.

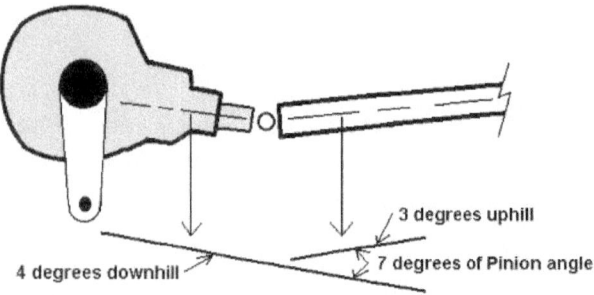

True Pinion angle is the difference between the angles of the pinion and the drive shaft.

4 degrees downhill / 3 degrees uphill / 7 degrees of Pinion angle

The pinion angle historically has been considered the angular difference between the sideview angle of the pinion shaft and the drive shaft. This angle, while needing to be minimal, has nothing in and of itself to do with correct driveline alignment. And, it does nothing to promote forward bite as sometimes spoken of. This angle must be the same and in the opposite direction at the transmission to drive shaft angle. Then the transmission output shaft and the pinion shaft will be inline and parallel.

Drive shaft angles are not only measured from a side view, but also from a top view. Some offset late model cars can have as much as a 1 ½ inch displacement of the rear of the drive shaft from the front. That equals almost 2 degrees of drive shaft angle at both the transmission and the pinion. So, we can align the drive shaft from a side view to zero angle and still have 2 degrees of drive shaft angle present in the system, which is plenty.

The latest recommendation might come as somewhat of a shock to most racers, but strictly for racing applications, near Zero Drive Shaft Angles are encouraged. Based on recent studies and testing, a race car with zero drive shaft angles at the transmission and pinion will not harm the u-joint bearings. The natural vibrations that are produced by the race car will cause the u-joint bearings to rotate enough to stay lubricated and not flat-spot.

In the real world, the drive shaft will never really maintain a zero angle configuration through the diving and rolling of the chassis as we lap the track. Starting at zero means that we will stay very close to zero at all times. For extreme applications, such as when racing at over 180 MPH at Daytona or a similar type of track, the teams will align the drive train for zero angles with the car placed at the on-track race attitude.

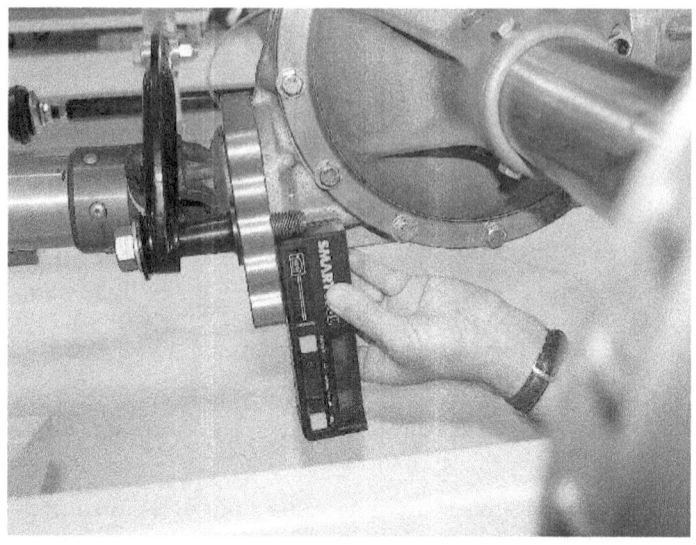

To achieve correct driveline alignment, we physically measure the pinion angle and the transmission angle. The first goal is to make those two parallel. Then we want to reduce the angle they make with the drive shaft to less than two degrees.

The mockup phase of car assembly is a good time to begin checking Driveshaft alignment. You have positioned the engine to meet all of its location constraints for best weight distribution and the rear axle location to meet the wheelbase requirements.

It is one of the things we do when going over our race car to create what we will call the total package that will be referred to a lot in the coming Lessons. There are certain "musts" we put on our list of things to do, and this is one of those.

NOTE: For overall consideration in drive line angle design, number one is to keep the angles equal and opposite at each end of the drive shaft (the transmission output shaft and pinion shaft are parallel) and to keep the driveline angles as small as possible.

Exam - In The Context Of This Lesson:

The Slip Yoke Does What?

1) Helps the tires to slip

2) Connects the pinion to the drive shaft.

3) Allows the drive shaft to turn without binding up

4) Attaches the drive shaft to the transmission

The Driveshaft Angles We Want Can Be Created From?

1) As a result of the motion of the rear end

2) From a top view angle due to miss-alignment of the engine and pinion shaft

3) From a sideview angle miss-alignment of the engine and pinion shaft

4) All of the above

Ideal Driveshaft Alignment Involves?

1) Having the driveshaft line up with the transmission output shaft

2) Installing the rear end with down angle in the pinion shaft

3) Making the transmission output shaft and the pinion shaft parallel

4) Mounting the engine/transmission and rear end so that there is less driveshaft angle

5) 3 and 4

Lesson Twenty – Types of Springs, How To Rate

Glossary:

Spring – A device that supports a load in either compression or extension. Springs can be a wound wire of different wire size and diameter of coils, a flat leaf design supported at each end and that bends to support weight, or a bar that is made to twist as load is applied – called a torsion bar.

Spring Rate – The actual rate of resistance to weight applied to it. We usually refer to a spring rate in pounds or kilos per inch. A 100 lb/in rate would deflect one inch if it were made to support a force of 100 pounds.

Spring Split – A difference in spring rates for the two springs on the same axle, or suspension system. Spring split is usually used in circle track racing where the cars only turn one way.

Motion Ratio – A difference in movement between one part of a suspension and another. It is usually used in reference to the wheels vertical movement in relation to the movement of the spring. When a spring is mounted on a lower control arm, it will move less than the wheel.

Force – The total pounds or kilos of work the spring does to support a measured load. A spring rated at 100 lb/in would develop a force of 400 pounds if compressed four inches. (100 pounds per inch rate compressed 4.0 inches; 100 x 4.0 = 400 pounds of force)

Weight – The actual measure of the loading on the tires measured in pounds or kilos of the parts of the car caused by gravity.

Sprung Mass – That part of a race car that is supported by the springs. The components that attach the sprung mass to the wheel structure are counted as part of the sprung mass using half their individual weight.

Over View – A car is supported by devices called springs. These come in many different varieties, but they all do basically the same thing. The provide a rising rate of force necessary to overcome first of all, gravity, and also the forces of aerodynamics, weight transfer and mechanical downforce due to cornering.

In modern day circle track racing, springs are being used like never before. Race teams now more than ever desire a low attitude and very little chassis movement. This provides very good aero qualities, a low center of gravity, and very little camber change, which the car likes. The result is a race car that is easier to setup and that runs faster.

In this Lesson, we will identify the different types of springs and tell you how they work to support the car and all of the loads related to performance driving. In developing a racing setup, we need to know the spring rates and we'll tell you how to find those rates accurately.

There is a sort of revolution going on in the industry today. Springs are being used in ways never before imagined. Combinations of springs, softer to stiffer springs, and stacked springs are being used in all forms of racing. So, this is not an ordinary lesson, it is the basis of much of what goes into race car design and setup.

The circle track teams are now learning what road racers have known for some time, that a stiffer setup that yields less chassis movement is a more stable package. What road race teams might now learn from circle track teams is that they could be running much lower to the track for better aero and a lower center of gravity. The circle track racers have perfected the low attitude setups.

Types of Springs:

Coil Springs – A coil spring is a bar that has been wound around a mandrel. The spring will have a designed wire diameter, an inside diameter of the mandrel, a specified height and a number of coils. The rate is determined by; 1) the hardness of the steel used to make the wire, 2) the wire diameter, 3) the diameter of the coils (used to find the length of the bar) and, 4) the number of coils (again related to finding the length of the bar).

The coil spring is by far the most common spring used in racing today. Other types less used are torsion bar systems, leaf springs, bevel springs, air springs, and bump stops and bump springs. Any way you look at it, springs support the car and dictate most of the dynamic properties of the race car.

Most coil springs are fairly linear in rate, meaning that as they support more load, the number of pounds or kilos it will support per inch or centimeter of travel stays consistent. For example, if I put 200 pounds of weight on a coil spring rated at 200 lbs per inch, then the spring will compress one inch (200 / 200 = 1.0")

We must know the rate of the springs we are using. This simple spring rater tells the team how many pounds (or kilos) a spring will support per inch (or centimeter) of spring travel. What we are getting into in the industry is also knowing the force a spring provides at a known distance of travel.

The if I add another 200 pounds to the weight supported by the spring, it should compress another inch for a total of 2.0 inches. If it were to compress more or less than an inch, it would be progressive (gaining rate), or digressive (losing rate).

It is common in some types of racing to use progressive springs to intentionally gain spring rate while the suspension travels. We almost never see digressive coil springs.

Torsion Bar Springs – Torsion bar springs are used on a variety of race cars, including every class from sprint cars to formula cars. These are simply coil springs that have never been wound. They work by being mounted solid at one end, having an arm attached to the other end, and a load being applied to the end of the arm causing the bar to twist. The resistance to twisting creates the spring rate.

Other types of springs include the torsion bar spring shown here on a sprint car. The bar is housed inside a tube (at the back end of the aluminum arm) and the spring rate is measured where the other end of the arm is attached to the axle tube.

This torsion bar spring, also on a sprint car, runs back from the front of the car to the straight axle tube. It rest on the tube instead of being bolted to it. The rate is the force at the point where the arm meets and rests on the axle tube.

The rate of a torsion bar is calculated using the following, being very similar to the coil spring; 1) the hardness of the steel used to make the torsion bar, 2) the diameter of the portion of the bar that twists, and 3) the length of the part of the bar that twists.

Gale Force Suspension offers a rating tool that can measure the sway bar rate and force while on the car. With so many different materials used to make sway bars, and the many different installation options we see, this rater takes the guessing out of sway bar rating.

To changes spring rates in a torsion bar system, you would simply install a bar with a different diameter, or change the length of the arm which changes the motion ratio causing the bar to twist more or less for the same amount of chassis movement.

Leaf Springs – Leaf springs are a flat bar, or layers of flat bars, that bend when load is applied to them. They have eyes at each end to mount them to the chassis and one end is mounted so as not to move and the other is mounted to a shackle and moves to compensate for becoming longer or shorter as the spring moves under load.

The leaf type of spring is common to older, vintage style, race cars. This one has a jack screw used to adjust the loading on the four tires. All leaf springs need a slider, or shackle, at one end to allow for length changes as the spring is loaded and un-loaded.

Leaf springs are a product of earlier production cars and trucks and are still used in some trucks and some race cars. The mono-leaf has become very popular for racing. This spring has one leaf, or flat bar, and can be designed to have different thicknesses, thicker closer to the axle and thinner out towards the eyes. Leaf springs are more common on dirt cars.

Half leaf designs may also be used. It is similar to a full leaf spring cut in half. In this configuration, the thicker part of the leaf is mounted solid to the chassis and the eye end is what supports the wheel assembly, usually the rear straight axle or independent suspension.

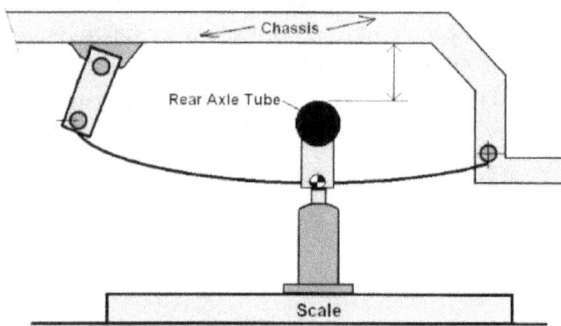

Race teams can rate the leaf spring while it is on the car. Rating leaf springs used to be nearly impossible, but now with this method, anyone with a scale can find the rate. Just measure the distance from the top of the axle tube to the frame. Jack the car us until the distance is less than it was by an inch or so. Then divide the number of pounds gained on the scale by the distance it moved closer to the frame and you will have a rate per inch.

The leaf spring is rated the same way a coil spring is. The rate is the number of pounds or kilos of load divided by the deflection in inches or centimeters to find the spring rate. These too are mostly linear in rate, although when using multiple leafs, the rate becomes progressive.

Bellows Springs – These springs are rarely used, but are available. The work by causing a deflection in a disc of specified material that resists being deflected. You usually stack a series of bellows springs together and the more discs in the stack, the softer the rate of the bellows springs.

The spring material could be anything from steel to carbon fiber. The uses include replacement of coil springs and use as a bump spring, which will be covered shortly. Just know that this type of spring exists and is not widely used.

Bump Stops and Bump Springs – The springs we have discussed above are referred to as ride springs. The car rides on these springs. There are supplemental springs that are used to add to the rate of the ride springs. These are bump stops and bump springs.

Bump stops and Bump Springs have been popular in the racing market place for some time. They are used in almost every form of racing from circle track to Formula One. They limit the travel of the suspension and provide a higher spring rate once a certain amount of travel has been reached. This shows a bump spring, a type of bump that provides an actual linear spring rate whereas most bump materials are linear in rate gaining a lot of rate in a very short distance of travel.

Both of these are usually mounted on the shock shaft of a coil-over spring (a spring mounted on a shock body). In some cases, the bumps are mounted separately from the coil-over. In any case, the bump device will add to the rate of the ride spring when it is "in-play".

For most applications, the term in-play means that the suspension has traveled far enough to engage the bump device. Then how much the suspension travels beyond that point determines the amount of force, or load, the bump supports.

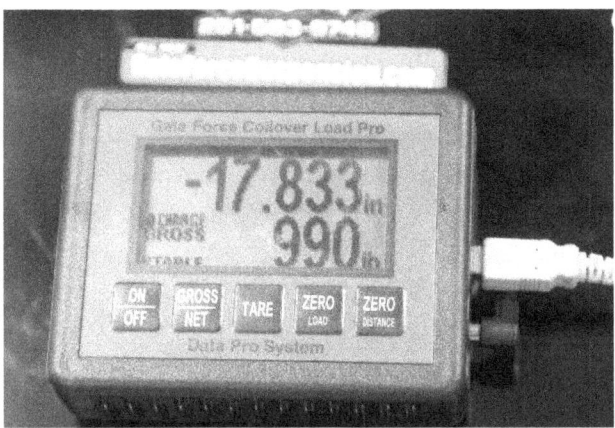

The newest and most useful information race teams are using in today's racing is Forced Based analysis. This means that teams are finding the force on the spring at a known distance of travel measured on the race track. This way they can know how much force is on the tire in order to re-arrange the corner heights to develop a better force distribution. We cover that in detail in Levels Two and Three of the RCT Course.

The bump stop is different than the bump spring in one important aspect. The stop is linear in rate, gaining rate at a very fast pace. The bump spring is mostly linear and gains rate at a consistent and predictable pace.

Bump devices are widely used in circle track racing, and prototype and formula racing. They provide a stable platform that can limit chassis travel, fix cambers and provide a much stiffer spring rate for roll control. There is a wide variety of rates and sizes available as well. The bump device is a major step forward in race car engineering for setups.

Air Springs – A fairly new and not yet popular spring is the air spring. It is the resistance to loading that comes from a cylinder that compresses air as it moves. The rate of the air spring is very non-linear and c can be easily raised or lowered by adjusting the air pressure in the cylinder.

The current uses of the air spring is for coil spring replacement and in use as a bump spring. Because is it so progressive, it may be hard to determine what rate to use in calculating roll angle and roll resistance. And temperature changes in the air can cause fluctuations in the spring rate.

Rating A Spring – We need to know the rate of a spring we are installing in our race car. It is the relationship of the spring rates that determines our setup balance. Here is how to rate a spring for a race car.

First, we must rate the spring in the range of motion it will be working within. A spring will very often provide a different rate as the suspension moves. This is not necessarily because the spring itself is progressive or digressive, it is usually because the spring angles, motion ratios, or bumps come into play and cause changes.

Carefully rating a spring is essential to having the correct setup. Most top race teams will rate their springs on a consistent basis and never install a spring that has not been rated. And always rate your springs within the range of travel they will experience while on the race car going around the race track.

When we rate a spring, we use a spring rating tool. There are many varieties on the market, but here is how they work. A spring is installed in the tool, and then compressed. The load is shown and recorded as the spring is compressed. We rate the spring by noting the gain in rate verses the amount of travel.

The first step in rating a spring is knowing the installed height of the spring and the total amount of travel the spring will encounter on the race track. This provides a defined range of motion and the area we need to know the rate within.

If a bump device is used in conjunction with the ride spring, then we can also rate the combination of ride spring and bump to find the new spring rate. No matter what the overall loading on the spring or spring/bump combo is, there is still a spring rate per inch or centimeter that we need to know for setup purposes. And, there is another way to look at this for further analysis.

Force Measurement – In modern racing, it is becoming popular and useful to measure the spring loading within the range of motion of the spring. This differs from spring rate in that it shows the force on the spring. This is used to compare loading to the companion spring at that end of the car.

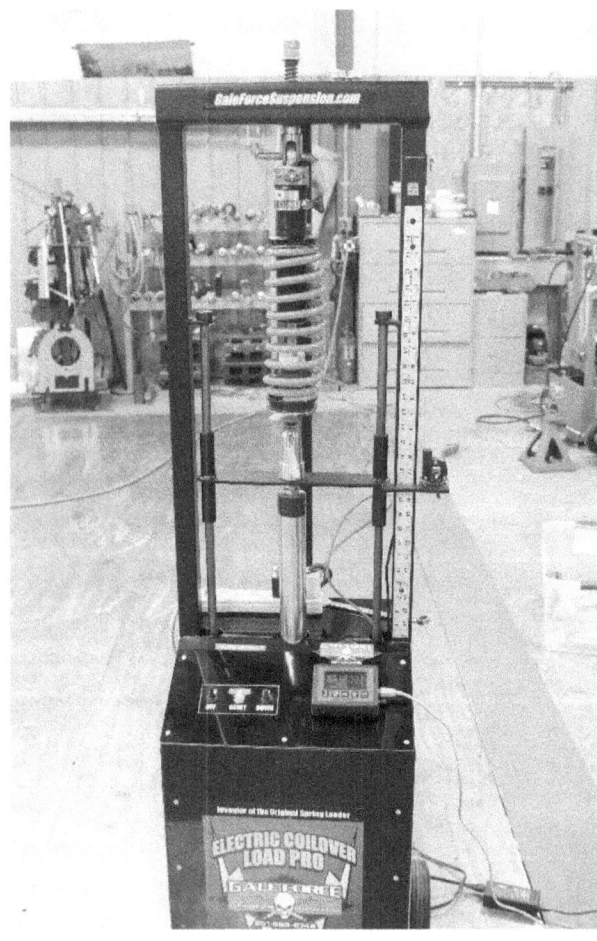

The Load Pro Coil-over force machine is the latest advanced tool for the racing industry. It records and prints out for small increments of travel the force on the spring. Knowing that and knowing the actual travel of the spring on the race track will give you valuable information about the force on the four tires in relation to each other.

The force on the spring has a direct relationship to loading on the tires. A way to balance a setup is to find a balance in the loading of the springs. This then translates to proper loading of the tires and can indicate problems with the setup where the tires are not properly loaded. More on that in future Lessons and Courses we will offer.

All of these different types of springs can, and are used in racing. The types can be mixed and matched depending on the overall goals for the type of race car and where it is intended to be raced.

Characteristics Of Springs In Setups – When we talked about setups, we talked about roll angles and how the front and rear must work together. The springs play a major role in this dynamic. Here are the ways a spring can and does influence the roll angle: 1) Spring rates combined per axle, 2) Spring split (i.e. spring rate on one side different than the rate on the other side), 3) Spring base (how wide apart the springs are mounted on a straight axle), Motion ratio (where the spring is mounted on a AA-arm lower arm).

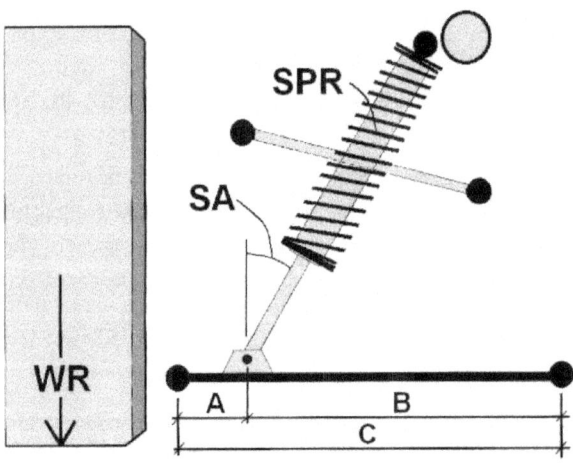

Each installed spring has a motion ratio of its motion to the wheel motion. The force it takes for the spring to support the wheel is much greater than the force of the wheel upon the ground. We can measure the force on the tire using scales. But on the race track we must use force based equipment to know what the tire contact patch loads are through the turns.

90

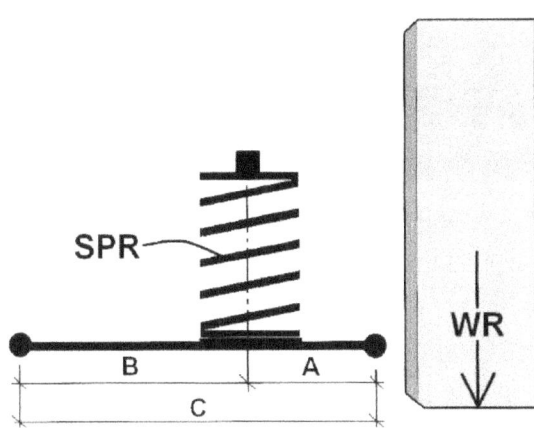

The "big spring" cars using more stock type of lower control arms and larger stock based springs must generate more spring force in order to develop the same wheel rate force. This is because the motion ratio is so much different and less than the coil-over sprung cars.

We are interested in finding the influence of the springs in creating the roll angle. This can be compared to roll stiffness, but there is more to the story than just that. As the chassis moves in dive and roll through the turns, the spring mounting geometry might change and that will affect the spring rate of the suspension. This gets somewhat complicated and we'll go over that part of springs in RCT Level Two.

Exam - In The Context Of This Lesson:

Springs Provide A Force To Overcome Which?

1) Gravity
2) Mechanical downforce
3) Aerodynamic downforce
4) Weight transfer
5) All of the above

A Spring That Is Linear Means?

1) It is straight and not bowed
2) Is mounted straight up
3) Is consistent in rate as it compresses
4) It is inline with the direction of travel

Coil And Torsion Bars Are Rated Based On?

1) Length
2) Wire or bar size
3) Hardness of the material
4) All of the above

A Bump Device Is Used To?

1) Smooth out the bumps in the race track
2) Overcome gravity
3) Add spring rate to the ride spring
4) Help the shocks overcome load changes

We Rate A Spring By Doing What?

1) Measure load changes in increments of travel
2) Using the range of motion of the spring
3) Measuring the travel of the spring on the track
4) All of the above

Springs Affect Setup Through?

1) How far they are mounted apart
2) Motion ratio
3) Changes in the spring rates
4) The difference in rates on a straight axle
5) All of the above

Spring Split Is?

1) When a spring is cut to make a different rate
2) A difference in spring angle for two springs
3) A difference in spring rate on the same axle
4) Different motion ratios for a AA-arm suspension

Lesson Twenty - One – Motion Ratio and Wheel Rate

Glossary:

Spring Rate – The actual rate of resistance to weight applied to it. We usually refer to a spring rate in pounds or kilos per inch. A 100 lb/in rate would deflect one inch if it were made to support a force of 100 pounds.

Motion Ratio – A difference in movement between one part of a suspension and another. It is usually used in reference to the wheels vertical movement in relation to the movement of the spring or shock. When a spring is mounted on a lower control arm, it will move less than the wheel.

Force – The total pounds or kilos of work the spring does to support a measured load. A spring rated at 100 lb/in would develop a force of 400 pounds if compressed four inches. (100 pounds per inch rate compressed 4.0 inches; 100 x 4.0 = 400 pounds of force)

Load – The actual measure of the weight on the tires measured in pounds or kilos of the parts of the car caused by gravity and other effects.

Over View – It is useless to know the rate of the springs used in our race car if we don't have other pertinent and important information. Remember our ultimate goal in race car setup is to provide the best load distribution on the four tire contact patches. In order to do this, we need to know how the loads on the springs end up on the tires.

We can calculate the wheel rate for a double A-arm suspension using simple formulas. We want to stress here that a solid axle suspension does not have a wheel rate. While it does in theory, that does not enter into consideration for calculating dynamic roll stiffness or roll angle analysis. We will cover this in much more detail in the RCT Level Two course.

So, for the purpose of this exercise, we are talking about wheel rate for a double A-arm suspension, or a swing arm rear link suspension, as you will see, or any other suspension with a motion ratio between motion of the spring and the wheel.

This photo shows how motion ratio works. In this rather complicated design, the pushrod (lower right quadrant) connects the spindle to a rocker arm (mid-photo). The rocker arm is then attached to the chassis forming a pivot point. Above this pivot point is the end of the coil-over shock mount. So, the wheel moves up, the push rod moves in, the rocker swings around and the spring compresses. The motion ratio is the ratio of the movement of the wheel to the amount the spring is compressed.

From the last Lesson, we can easily determine the loading on the springs if we know the range of motion the springs operate within. We just compress them to those ranges and then record the loads. But it doesn't stop there, that loading is transferred to the wheel and tire. How it gets there is what we will learn now.

Motion Ratio – In a AA-arm suspension, the spring is almost never mounted to move on a one-to-one ratio with the wheel movement. In almost every case, the spring is mounted so that it moves less than the wheel and this is called a motion ratio because we can calculate the ratio.

If, on a AA-arm suspension, the spring were mounted one-quarter of the way from the ball joint to the chassis mounts, then we would have a motion ratio of 0.75:1, the spring moves 0.75" and the wheel moves 1.0". That is 75 percent movement of the spring to the movement of the ball joint, or wheel. Pretty simple, right? Not so fast.

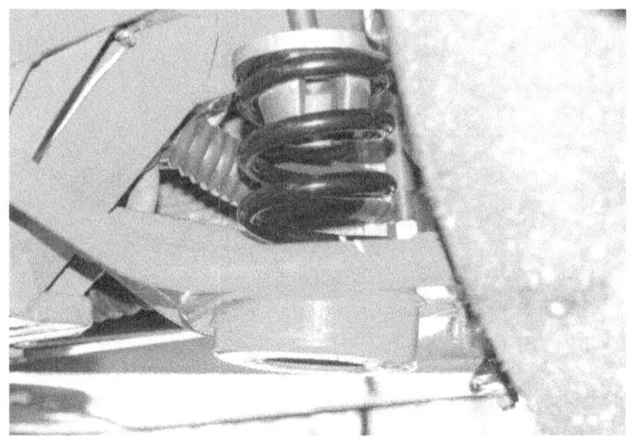

In this example we see a stock type of spring mounted in a bucket designed into the lower control arm. The wheel moves more than the spring in vertical motion. It is the division of the spring movement by the amount the wheel moves that is the motion ratio.

Motion ratios are not limited to double A-arm suspensions. Some race cars have rear suspension parts and links that create a motion ratio that affects the spring rate the car utilizes, or "feels". In this example, we see a three or Z-link rear suspension link going forward from the straight axle tube. The coil-over spring is mounted onto the link some distance back from the point where it attaches to the axle tube. When the chassis moves vertically, the spring moves less than the chassis. The spring may also be mounted at an angle to the link which creates another motion ratio.

Now we have the motion ratio of the movement of the ball joint to the spring mount. But, in many cases with coil-over springs especially, the spring is not mounted at ninety degrees to the control arm. Now as the spring compresses, its motion is not a one-to-one motion, it is yet another motion ratio.

If the spring were mounted at 20 degrees, then as the bottom of the spring moved up, the spring would compress 94.0 percent of the amount the bottom moved. So, our combined motion ratio in this AA-arm suspension system is 94 percent times 75 percent which looks like this. We take 0.94 times 0.75 = 0.705. This means that our spring compresses 70.5 percent of the distance the wheel moves.

In a straight axle suspension system, there is a motion ratio of the springs to the wheels, but that motion ratio is not useful in dynamic calculations. This is because the dynamics of the straight axle tell us that spring base determines roll resistance. This very important to know now because we will discuss this relationship in future Lessons.

Wheel Rate – We use the motion ratio in a AA-arm suspension to determine the wheel rate, or spring rate at the wheel. This is something we definitely need to know in order to plan out our setups. The way to find the wheel rate knowing the spring rate is simple. It is the combined motion ratio (mount motion ratio plus spring angle ratio) squared (motion ratio number times itself).

If we know the overall motion ratio, which combines the control arm motion ratio plus the spring angle motion ratio, we just square that number. In our example above, we had a combined motion ratio of 0.705. If we square that, we get 0.497, or 49.7 percent. The wheel rate ends up being about half the installed spring rate.

Taking a spring rate of say, 300 lb/in (pounds per inch), times 0.497, and we get a wheel rate of 149.1 lb/in. If we know the wheel travel, we can multiply the wheel travel times the wheel rate to find the tire loading. In this example, let's say the wheel travel were 4.0 inches.

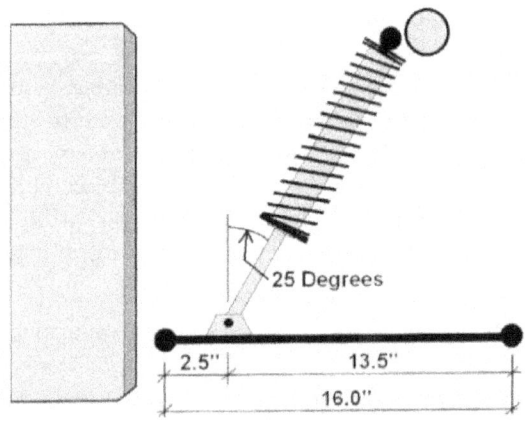

This is a typical coil-over mount we see on dirt or asphalt race cars. The bottom of the coil-over is mounted to the lower control arm some distance from the ball joint and the top of the coil-over is mounted to the chassis. There is a motion ratio created by the mounting points and another motion ratio created by the shock angle.

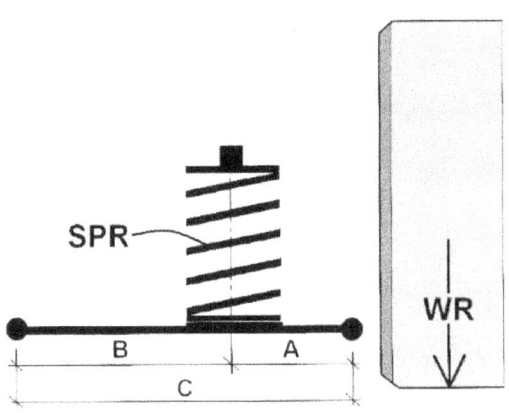

In the big spring type of suspension, the spring is mounted mostly at right angles to the lower control arm. So, in this case, we only have the actual mounting motion ratio to calculate to find the wheel rate.

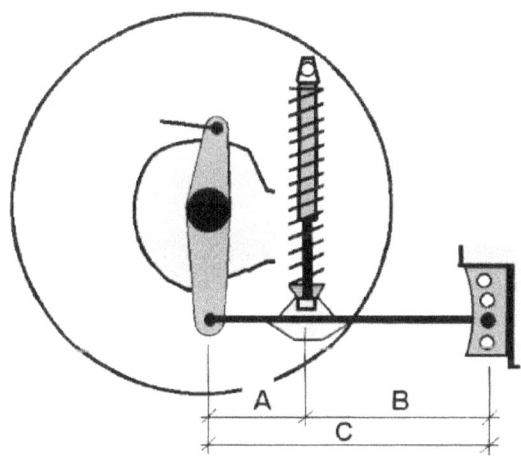

This is a more detailed sketch of a rear link where the spring is mounted on the link. We can see where as the chassis moves vertically, the spring will move a smaller amount. To visualize the "wheel rate" so to speak, we can imagine this to be a double A-arm setup with the axle mount acting much like the wheel. Then it is just a matter of doing the math.

If we multiply the wheel rate of 149.1 times the wheel travel of 4.0, we get an additional wheel loading of 596.4 pounds over that which existed at static ride height. Remember, we already had a wheel load holding the car up that was resisting gravity before we traveled the 4.0 inches.

This number, 596.4 lb, represents the additional load on the tire at that wheel travel. If we add the static loading, which can be directly measured with a scale, of say 550 pounds, this extra loading is added to that for a total loading of 1146.4 pounds at mid-turn.

How To Easily Measure Motion Ratio – Measuring the lengths of the control arms and the distance from the ball joint to the spring mount takes time and can be inaccurate. And, measuring the shock angle is difficult as well and can also be inaccurate. There is a much more simple way to find your motion ratio.

To find the motion ratio the simple way, start by supporting the chassis at nearly the normal ride height. Remove the spring or coil-over. Then measure the length of the spring mount. For a coil-over, just measure the centers of where the ends are bolted. For a stock type of big spring car, measure from the bottom of the bucket in the control arm up to a point directly above the bottom point.

Now place a block of wood or metal that is close to the amount the chassis will travel on the race track under the tire. Note how thick the spacer under the tire is. Then remeasure the spring length. Divide the amount the spring traveled by the spacer thickness and you have the motion ratio. To find the wheel rate, square the motion ratio and multiply that times the spring rate.

Motion Ratio and Wheel Rate are important concepts to understand. We don't need to go into detail about how they are ultimately used in race car dynamics, we'll get into that in RCT Level Two. Now that you have been introduced to this, some of what we will be getting into later on will make more sense.

Exam - In The Context Of This Lesson:

In A AA-arm Suspension, The Spring Moves?

1) Separately from the wheel
2) A different distance than the wheel
3) More than the wheel
4) Half the distance of the wheel

Motion Ratio Is Defined As?

1) How far the spring moves
2) How far the wheel moves
3) Relationship of spring movement to wheel movement
4) How far the chassis rolls

Wheel Rate Is What?

1) The distance the wheel moves in the turns
2) How much load is on the tire
3) The shock rate times the motion ratio
4) Motion ratio squared times the spring rate

Mid-turn Tire Loading Is?

1) The gravity loading plus spring rate
2) Gravity loading plus wheel travel loading
3) A product of motion ratio
4) More than the spring loading

Lesson Twenty-Two – Spring Base and Choosing Springs

Glossary:

Spring Rate – The actual rate of resistance to weight applied to it. We usually refer to a spring rate in pounds or kilos per inch. A 100 lb/in rate would deflect one inch if it were made to support a force of 100 pounds.

Spring Split – A difference in spring rates for the two springs on the same axle, or suspension system. Spring split is usually used in circle track racing where the cars only turn one way.

Motion Ratio – A difference in movement between one part of a suspension and another. It is usually used in reference to the wheels vertical movement in relation to the movement of the spring. When a spring is mounted on a lower control arm, it will move less than the wheel.

Force – The total pounds or kilos of work the spring does to support a measured load. A spring rated at 100 lb/in would develop a force of 400 pounds if compressed four inches. (100 pounds per inch rate compressed 4.0 inches; 100 x 4.0 = 400 pounds of force)

Load – The actual measure of the weight on the tires measured in pounds or kilos of the parts of the car caused by gravity and other effects.

Over View – As covered in the previous Lesson, the spring base for a AA-arm suspension is the width of the tires. This is because the spring rate is transferred out to the wheel contact patches. It is as if the car were riding directly on the springs.

For the straight axle suspension system, it is a little different. The chassis "feels" the spring base in those systems as the width of the springs, or more appropriately, the width at the top of the springs. So, here is how that looks.

Our choice of spring rates will determine, in part, our roll stiffness and roll angle for a suspension system. It is the calculation of roll angles that is at the heart of setup for any race car. In the past, the straight axle suspension system was looked at differently and the spring rates were translated out to the wheels. This is not the correct way to look at the vehicle dynamics of this suspension system, and I say that having proof developed by years of testing.

For a double A-arm suspension, the spring base is the width between the centers of the tires. For a solid axle suspension, it is different. The chassis "feels" the spring base at the point where the chassis rests on the springs. For a coil-over design, the top mounting point is the top of the spring, so to speak. It is the distance between the top of the two springs that forms the spring base for the solid axle suspension.

Spring Base – The spring base is the width of the springs where the chassis mounts to them in a straight axle suspension system. For a AA-arm suspension the spring base is a little different. Because we translate the spring rate out to, basically the ball joint or end of the lower control arm, that ball joint becomes our spring as such. So, the spring base is the width of the distance between the centers of the two lower control arm ball joints, one on each side.

This concept of spring base is important to understand because the roll stiffness and ultimate roll angle is a product of: 1) the spring base width, 2) the moment arm length, 3) the spring rates, 4) the magnitude of the lateral force, 5) spring split. It is the last of these that we need to explain a little more about.

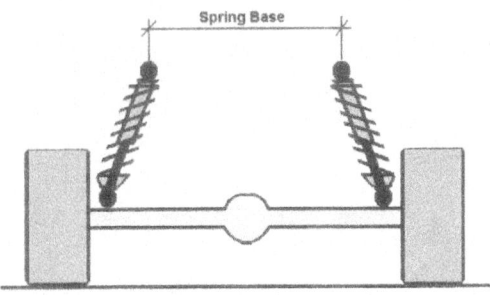

This illustrates the spring base for a solid axle suspension. The base is located where the chassis attaches to the springs, or coil-over assembly. This sketch closely resembles a dirt late model configuration with a lot of spring angle. The car does not know where the tires are and the width of the tires has no effect on the dynamics of the suspension. It's as if the chassis were sitting on springs attached to the ground. And the wider the spring base, the greater the roll resistance, or stiffness of the chassis.

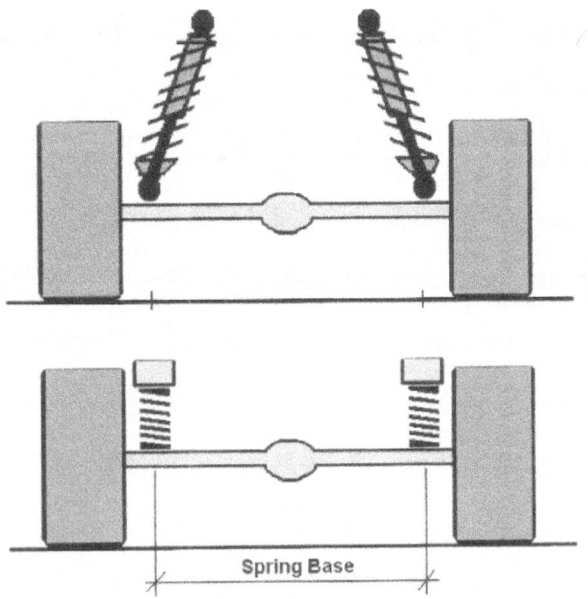

This illustrates the two basic designs for a straight axle suspension. Some designs place the springs on top of the axle tubes and use stock type of springs, some mount to truck arms in a similar way (top series Cup cars use this design) and some are the coil-over design we already talked about. Note that the spring base for the stock spring mounted on the axle tube is wider than the coil-over design used on dirt cars which have a lot of spring angle to them. Asphalt late model cars incorporate coil-over springs that are much more vertical and therefore have a wider spring base than what is shown.

Spring Split - The spring split mentioned in 5) above means that for a circle track stock car, we can and often do, install springs of different rates on each side of the car. The RF spring might be stiffer in rate than the LF spring. And the RR spring might be stiffer than the LR spring. This is almost never the case in road racing where the dynamics must be the same for left and right hand turns.

The benefit to using spring split is this. When we are turning and especially when the turn is banked, the forces on the car will do two things. They will try to roll the car over, and they will try to compress the car. The combination of the two forces, gravity and lateral force, combine for a net force that has a magnitude and direction that will roll, as well as compress, the chassis. The compression is called mechanical downforce.

It is this compression that plays into spring split. When we compress a suspension that has different spring rates on each side, we create a roll angle. The softer spring compresses more than the stiffer spring. So, in a left turn on a high banked track, if the LR spring were less rate than the RR spring, the compression would cause a roll angle to the left, or what we can call a negative roll angle.

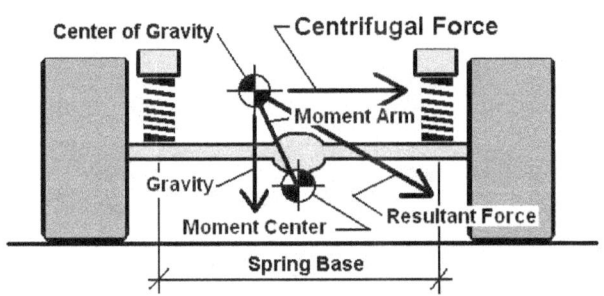

The dynamics of the straight axle suspension are pictured here. We have a lateral force called the Centrifugal force that is a result of traveling on a arc at high speeds. We also have Gravity which is always present. When we combine these two forces, we get a Resultant Force. Note the angle of this force. One a banked track, it will be pointed well between the two tires and provides what we call mechanical downforce pulling down on the chassis. With spring split we first talked about in Lesson Twenty, we can create more or less roll angle with this type of suspension easily.

If the lateral force were creating a positive roll angle to the right in a left turn, then we would add the two angles, and if let's say the spring split angle were -1.0 degrees and the lateral force roll angle were +2.0, the result would be a net roll angle of +1.0 degrees. This really does happen in a race car using spring split.

The spring split concept was developed to counter the roll forces for a straight axle rear suspension. It became hard to limit the rear roll angle as much as we could limit the front roll angle in a stock car. In the AA-arm suspension, we can install larger sway bars, stiffer springs, and/or bumps to reduce the roll angle to very low numbers.

In order for the rear suspension to be matched to the front for roll angle, we cannot, or should not install very stiff springs. Raising the height of the rear moment center can reduce roll angle, but there are limits to how high we can and should go.

The answer to the problem was spring split. Race teams found that by installing springs whereby the inside spring were softer than the outside spring, the roll angle was significantly reduced. By varying the amount of split, the team could effectively tune the roll angle balance of the chassis and create the perfect setup.

Choosing Circle Track Spring Rates – So far we have been talking mostly about stock cars and straight axle suspensions, but choosing spring rates is a topic of discussion meant for any race car, no matter the design. So, we have split the discussion into circle track and road racing sections.

We choose our spring rates for circle track racing with the dynamic balance and roll angle harmony as the primary considerations. And, for some aero sensitive cars, the rake of the chassis and body has a lot of importance. But the primary goal is dynamic balance.

For CT racing, we have more tools in the tool box than for other types of racing. We have, as discussed above, spring split. We also have a variable on roll center design and camber layout we can use. Whereas the road racing cars need to be more symmetric in their design left to right, the CT car does not.

The circle track cars can run different cambers left to right, different roll center location than being centered in the chassis, and they can run different spring rates on each side. So, it then becomes almost too easy to balance a CT car for roll angles. You just have to know what to install for spring rates, etc., the tools are all there. We only need satisfy one direction of turning.

The other thing we can change to affect the roll angles is the sway bar. And in CT racing, we can either have just a front sway bar, or a front and a rear. For rules that limit rear spring split, as has happened in the past, the team can install a rear sway bar to help limit rear roll.

Road Racing Spring Selection – For road racing cars, including stock type cars, prototypes and formula cars, the springs must be symmetric in rates. These cars must turn in both directions and so any spring split could work one way, but not the other. Its effect would be reversed when turning the other way.

Another major consideration for road racing cars is pitch attitude. Many of these cars are very aero sensitive. The angle of the chassis fore to aft, is critical and any change in pitch angle can have significant effect on the amount of downforce and downforce balance front to rear. This can cause a huge change in front to rear grip and handling balance.

To produce the ideal pitch angle and to help maintain that pitch, the teams will choose spring rates and bumps that will put the chassis at the correct pitch. This is sometimes done at the expense of dynamic balance, or as we call it, matching of the roll angles. This is where the sway bar, or anti-roll bar comes into play.

The sway bar can make up for deficiencies in the spring stiffness to help match the race cars roll angles. Suppose that the rear suspension was too soft and was rolling too much and more than the front. If the front angle seemed reasonable, then the team could increase the rear sway bar stiffness to more evenly match the front and rear roll stiffness and ultimately the roll angles.

In modern prototype cars and formula cars, there are methods that are used to adjust the stiffness of the bar. If the bar arms are made of bendable blades, then these blades can be rotated. The range is from ninety degrees to the attachment link (flat and producing the least resistance to bending) to zero degrees, or inline to where there is no bending for full stiff positioning.

The bar itself can be switched out to a smaller or larger bar, whichever is desired. In our example, we would install a larger diameter bar to try to work within the adjustment range of the blades if full stiff were not enough.

The selection of spring rates has to be done in conjunction with other settings and parts on the car. The spring rates are just one part of the dynamic mix that controls the roll angles at the front and rear of the car.

Exam - In The Context Of This Lesson:

A Straight Axle Spring Base Is?
1) The bottom of the springs
2) The width of the tire contact patches
3) The distance between the top of the springs
4) The spring rates times the track width

The Roll Angle Is A Product Of Which Of The Following?
1) the spring rates and spring base
2) the moment arm length,
3) the magnitude of the lateral force,
4) spring split
5) All of the above

Spring Split With Inside Spring Softer Causes?
1) Less roll angle
2) More roll angle
3) The car to be tight
4) The car to be loose

Circle Track Race Cars Can Do What That Road Racing Cars Can't?
1) Handle high banked turns
2) Run different rate springs side to side
3) Run softer rear springs
4) Run a stiffer overall spring rate

Road Racing Cars Can Best Adjust Roll Angles By?
1) Using different spring rates at each end
2) Making the spring stiffness more or less
3) Making changes to the sway bar stiffness
4) Moving the moment center side to side

Lesson Twenty-Three – What Shocks Do and Don't Do

Glossary:

Load – The actual measure of the weight on the tires measured in pounds or kilos of the parts of the car caused by gravity and other effects.

Compression – The motion of the shock when the shaft is pushed in and the shock becomes shorter.

Rebound – The motion of the shock when the shaft is pulled out and the shock becomes longer.

In Transition – A car that is either slowing or accelerating. Transition portions of the race track are the entry during braking and the exit during acceleration.

Over View – A shock is a control and timing device. It works in conjunction with the springs and other components to control the motion of the springs and to regulate the distribution of loading on the tire contract patches. There are things a shock can do, and things it cannot do. Here is a list of those:

What Shocks Do Not Do – 1) Shocks do not and cannot support the car, i.e. they are not springs, 2) Shocks do not increase or decrease the amount of load transfer, 3) Shocks do not influence chassis dynamics if they are not moving, 4) Shocks are not a cure-all for basic handling problems.

What Shocks Actually Do – 1) Shocks control the speed of the vertical motion of the part of the suspension they are bolted to by controlling the speed of the rebound and compression of the spring, 2) Shocks with varying designs of resistance allow more or less rapid movement of a suspension corner than opposing corners, 3) Shocks regulate the amount of time it takes for a corner of the car, while in transition, to assume a new ride height, but load transfer happens immediately, 4) Shocks can be used to redistribute the loads at the four corners of the car as the car is in transition on corner entry and exit, 5) and, shocks can be used to overcome the ride springs on one or more corners of the car when running bump setups.

This road race car is at mid-turn and has stopped slowing and not yet started to accelerate off the turn. At this point on the race track, the shocks are doing no work. Shock do not have a dynamic function if they are not moving. This is the first important principle you must learn and know about racing shocks.

This Pro Late model race car is entering a turn and is under braking. The shocks are managing the transfer of weight from the rear to the front as a by-product of rapid slowing. The rear shocks are managing the rebounding of the rear springs and the front shocks are managing the compression of the front springs. This is where shocks can make a difference in the dynamics of the chassis.

Let's go over these one at a time. It is thought by some that a shock can help tune the setups of the car at mid-turn, steady state conditions where the car is neither slowing or accelerating. That is not true. Shocks may help or hurt how you got to mid-turn, but once there, shock no longer play a part.

The reason a shock cannot do that is because of a simple fact about shocks. The do not do any work unless they are moving. A spring works all the time it is loaded. Even if it is not moving, a spring is working providing force that holds up the car. Not so with a shock. A shock does no work when it is not moving.

When a weight is put onto a spring and it has to compress, it resists that motion. When it releases as load comes off of it, it wants to release that energy quickly and move very quickly. A spring resist compression motion and promotes rebound motion. The shock, more than anything, resist the rebound motion of the spring. In a much smaller way, it can control the compression of the spring.

When the race car is at steady state condition, meaning it is not accelerating nor slowing, the shocks have no influence on the distribution of loads on the four tires. It may as well be sitting on the trailer. One other thing about this photo is the angle of the shock. It will move at a slower speed than the spring as the chassis moves because of the extreme angle. Therefore, its rate must be higher than if it were mounted straight up in order to do the same work.

The stiffer the shock is, the slower the motion of the spring is, in both compression and rebound. If we install shocks of different rates of compression and rebound at the four corners of the car in relation to the spring rates, we can cause the loads on the four wheels to change while the car is in transition.

At two points on the race track, the weights on the four tires are steady, on the straights and in the turns at mid-turn, steady state (not slowing, not accelerating). The weight has transferred and has settled onto the springs and tires.

We setup our cars so that the grip generated load distribution is such that the grip at the front and rear are the same. But when the car is in transition, we might need a different load distribution. Since the shocks work while in motion, and when braking and accelerating the suspension is in motion, the shocks can help us out.

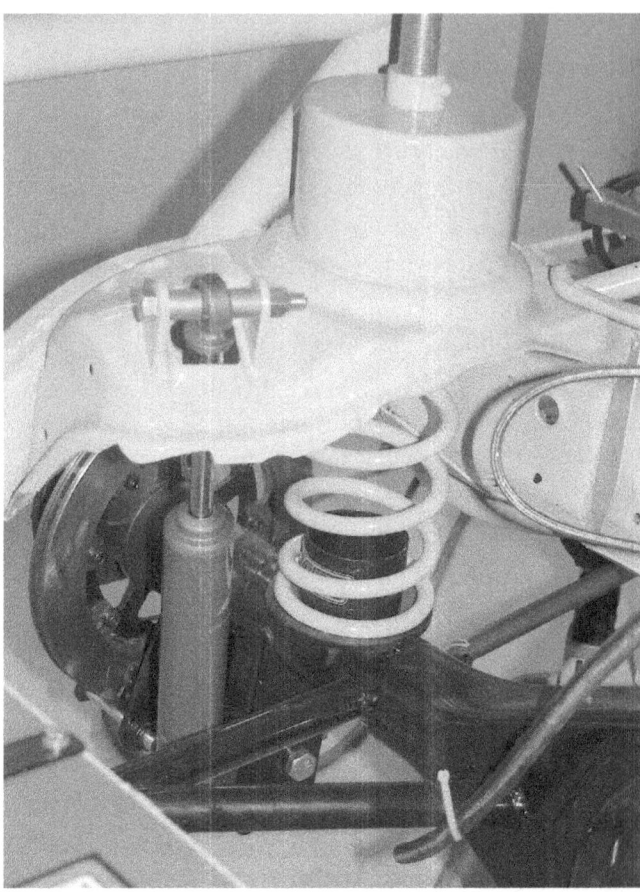

This shock is mounted where it will do the most good along with the motion of the spring. It is mounted vertically and in nearly the same distance from the centerline of the car. Its motion ratio is nearly one to one.

By altering the amount of control the shocks have in either compression or rebound, we can control the load distribution during those times when the suspension is in motion.

When we add load or take load off a spring, it wants to change to a new height. This is what happens on the corners of the car as weight transfers during braking, cornering, and acceleration. A shock controls that motion and dictates how long it takes for the spring to assume the new height.

This is the essence of what a shock does. It is up to us to design our shock package to time the movement of the springs and suspension so that the load distribution is what is needed for maximum performance. We will cover how race teams do that in RCT Level Two.

Exam - In The Context Of This Lesson:

What A Shock Does Not Do?
1) Support the race car
2) Add load to the tires
3) Promote load transfer
4) Change the handling at mid-turn
5) All of the above

What A Shock Does Do?
1) Adds load to the rear tires
2) Regulates amount of load transfer
3) Resists motion of the springs and suspension
4) Tunes mid-turn handling

Shocks Redistribute Loading On The Tires By?
1) Promoting faster or slower movement of each spring
2) Allowing quicker load transfer
3) Providing more compression than rebound resistance
4) Causing changes in the handling

The Suspension Is In Steady State When?
1) On turn entry
2) On turn exit
3) At mid-turn
4) Half way down the straight away
5) 3 and 4

A Spring Moves To A New Height When?
1) The shock is stiffer
2) Weight is added
3) The shock is softer
4) When weight is removed
5) 2 and 4

Lesson Twenty Four – How A Shock Works

Glossary:

Load – The actual measure of the weight on the tires measured in pounds or kilos of the parts of the car caused by gravity and other effects.

Shock – A device that controls spring movement. This is also called a dampener, but the terms are interchangeable.

Twin Tube Shock – A shock design that uses two tubes, one to hold a gas bag to separate the shock oil from the gas, and one to hold the shock piston, oil and shaft.

Gas Pressure Shock – A shock that uses pressurization of the gas to separate the oil from the gas.

Shock Body - The tube that holds the fluids and other parts including the shaft, separator piston and valve attached to the end of the shaft. The end of the shock body is attached to the chassis or control arm, depending on which way the shock is mounted.

Separator Piston – A free floating piston that is used to separate the shock oil from the pressurized gas.

Shock Piston – A part that is attached to the end of the piston shaft and to which is attached valve discs that regulate the flow of shock oil as the shock moves in compression and rebound.

Shock Valve – A series of discs of different thicknesses and diameters that are mounted on the shock piston that control the amount of fluid that passes through it when the shock moves in compression or rebound.

Bleed – A small hole, or needle valve, that allows the free passage of oil from one side of the shock piston to the other.

Over View – A shock works to control movement of the springs. It does this by making a fluid move through openings of various sizes and it takes longer and more effort for the fluid to move through a smaller opening than a larger one. That is the short explanation. Here is a longer one that will help you completely understand how they really work and how we can change how efficiently they work.

How Shocks Work - Shocks resist motion by using a piston that must move through a fluid (thin oil) as the suspension moves so that the fluid must pass through holes, valves and slots. Varying resistance is created when the oil is forced through different sized openings. The resistance is usually different for each of the two motions, compression (inward motion) and rebound (outward motion).

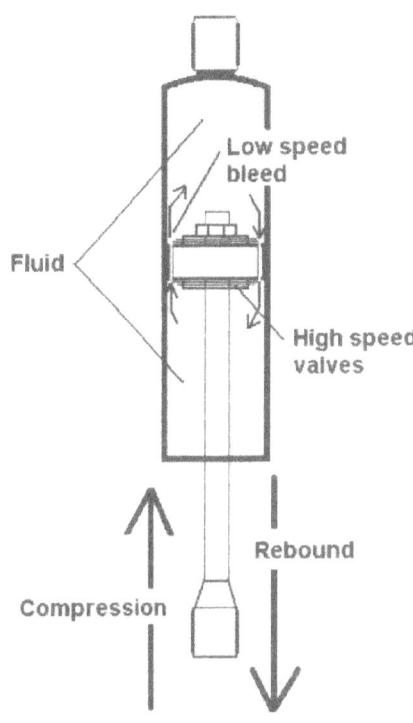

THE INSIDE OF A RACING SHOCK

This is basically how a shock works. It takes energy and force to move a shock because when it moves either in or out, fluid must pass through holes and ports. The smaller the holes and valve ports, the harder it is for the shock to move. So, the larger holes and valve ports will allow more rapid movement of the shock. And smaller holes and valve ports will slow the movement, or require more force to move it the same speed. In this way we can control different spring rates.

All racing shocks are of two basic designs – twin tube and mono-tube and can be either gas pressurized or "low" pressure. The twin tube has literally two tubes, the inside tube is where the work is done and the outside tube is a reservoir that holds extra fluids and a flexible gas filled bag.

The mono-tube design is almost always pressurized to some PSI (pounds per square inch pressure). Most every dedicated race car now uses the mono-tube, gas pressure shock.

Shaft Displacement – An important feature related to the design of a racing shock is called shaft displacement. When the shock shaft is pushed into the shock body and into the fluid, it takes up space. Suppose we pull the shock shaft out as far as it would go, fill the shock body with oil only, and then reseal the shock body. If we tried to push the shaft into the shock body and into the volume of oil it would not go because the shaft would be trying to compress and displace some of the oil and oil will not compress.

We need a space inside the shock filled with a substance that will compress. Gases such as air and nitrogen will compress. Every shock needs to have a certain volume of gas, usually nitrogen, along with the fluid, in order to allow for the shaft to move into the shock body and take up space. Nitrogen is a dry gas that suits our purpose and is widely used in racing shocks.

Keeping the Oil and Gas Separate – The Nitrogen gas we put into our shocks to compress and allow for the volume of the shock shaft must be separated from the fluids in all shocks. The gas can be contained inside a plastic bag or separated by another piston.

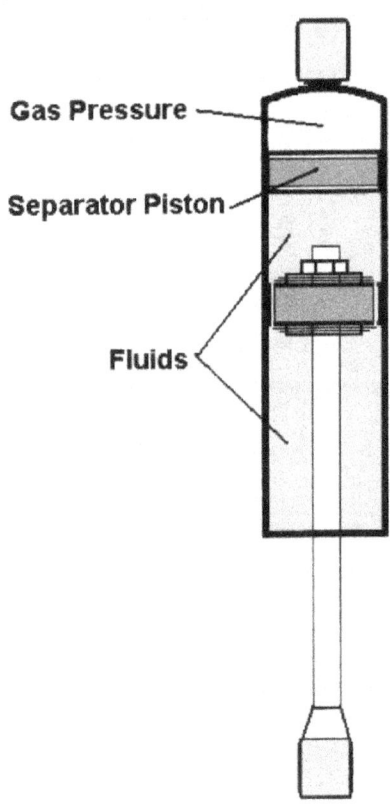

In reality, every shock must have a chamber that contains a gas. Gas is compressible whereas fluid is not. When the shaft is pushed into the shock body, it takes up space and displaces shock fluid. There must be a way to allow the volume of the shock shaft to exist within the shock body. That way is by way of gas compression. As the shaft moves in and takes up space, the gas compresses to provide the room for the shock shaft. A piston separates the gas from the shock fluid.

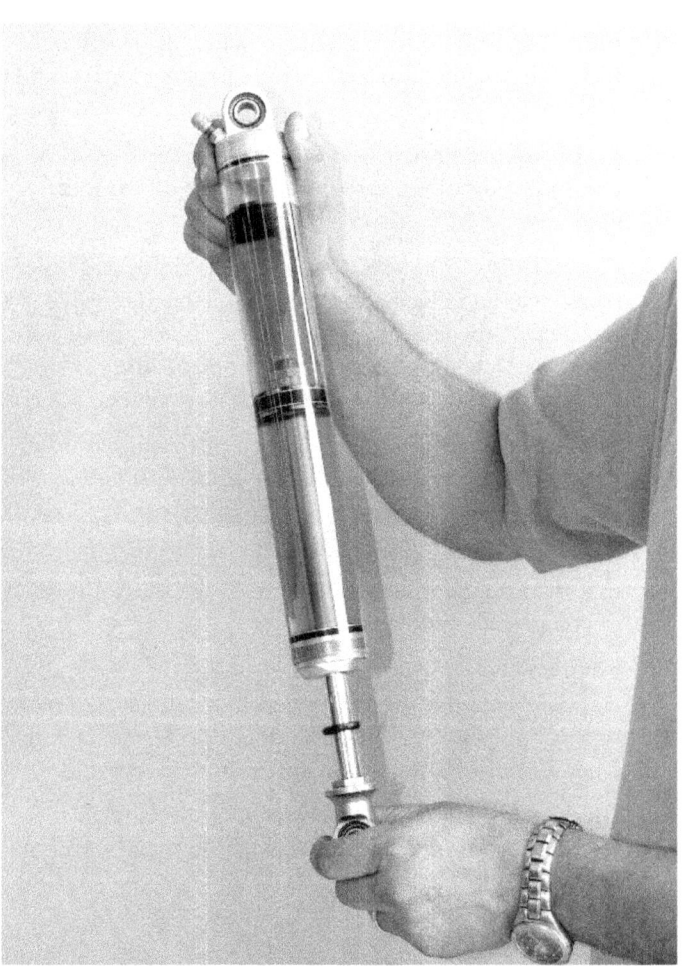

This is a demonstration shock with a clear body showing the internal parts of a gas pressure shock. Here we can clearly see the red shock fluid, the shock piston that controls the flow of fluid in compression and rebound, and the separator piston between the red shock fluid and the clear compressible gas at the top.

Compression and Rebound - The compression control side of what the shock does resists: 1) the bump movement of a corner of the car when we hit bumps (rises in the track), 2) the bump movement of the front shocks due to the transfer of load to the front end during braking/deceleration, 3) the bump movement due to the transfer of load to the rear upon acceleration, 4) and the speed at which the right side springs compress when the lateral forces are applied as we deviate from a straight line and turn into a corner.

The rebound control side of the shock resists the following: 1) rear chassis rebound, or vertical movement, when decelerating, 2) front chassis rebound movement upon acceleration, and 3) left side

rebound vertical movement caused by load transfer from the inside to the outside of the turn as we negotiate the turns.

Here we see a closeup of the shock piston and the valves lying under the nut. There are different designs of pistons that react to the movement of the shock differently. The primary two are the Digressive and Linear piston designs. Along with those different pistons, we have many different thicknesses of valve discs that produce many different levels of resistance force to control different rates of springs.

Because the bags used in twin-tube designs are not pressurized, these shocks are referred to as non-pressurized shocks. In truth, as the shock shaft is pushed into the shock body, some amount of pressurization must take place due to the displacement of the shaft and the compressing of the gas inside the body.

In mono-tube shocks, separation is accomplished by installing a second separator floating piston, which provides a seal that separates the fluid from the gases. A Schrader valve that is installed in the shock body at the gas chamber end of the shock allows us to pressurize the gas inside the shock. This pressure ensures that the gas will be forced to separate from the fluids at all times. That is because the seal on the piston will usually allow the gas to seep past the piston from the fluid side to the gas chamber, but seals the heavier fluid from escaping into the gas chamber.

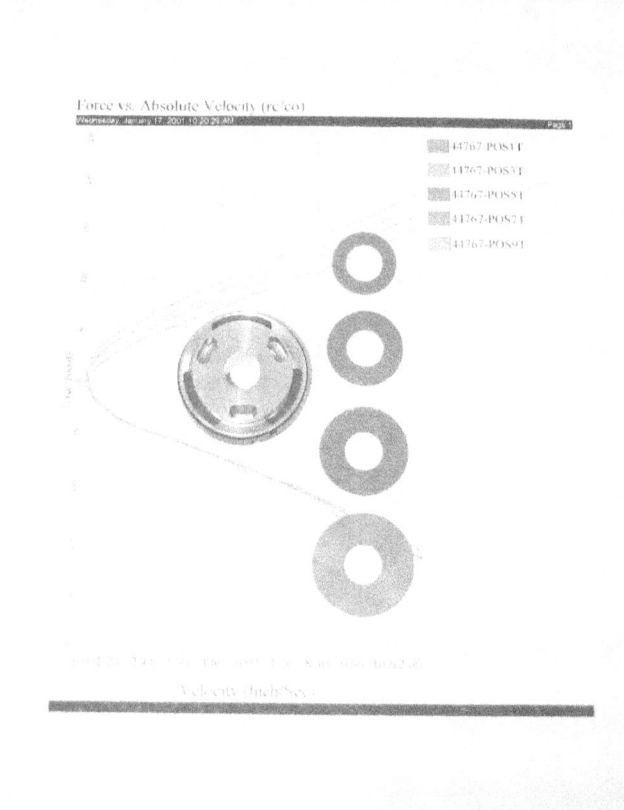

The shock valve and the valve discs are shown here. Fluid flows through the slots in the piston and must force the valves open in order to pass through. We can stack different numbers of discs and use different thicknesses of discs to create the resistance force we need.

The amount of resistance that the shock provides with each movement, compression and rebound increases with the speed at which the shock is forced to move. Low speeds create low resistance and high speed movement creates higher resistance.

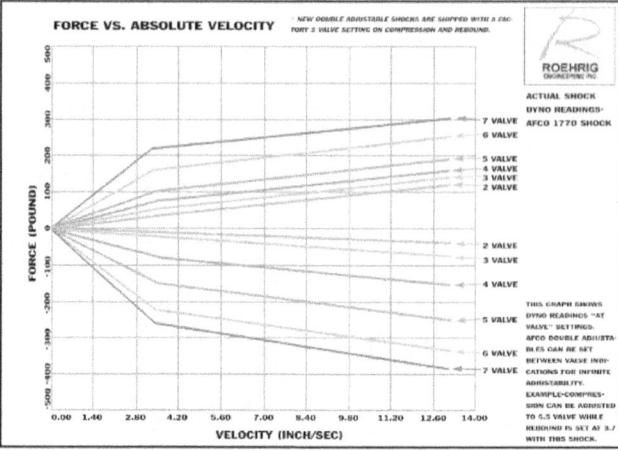

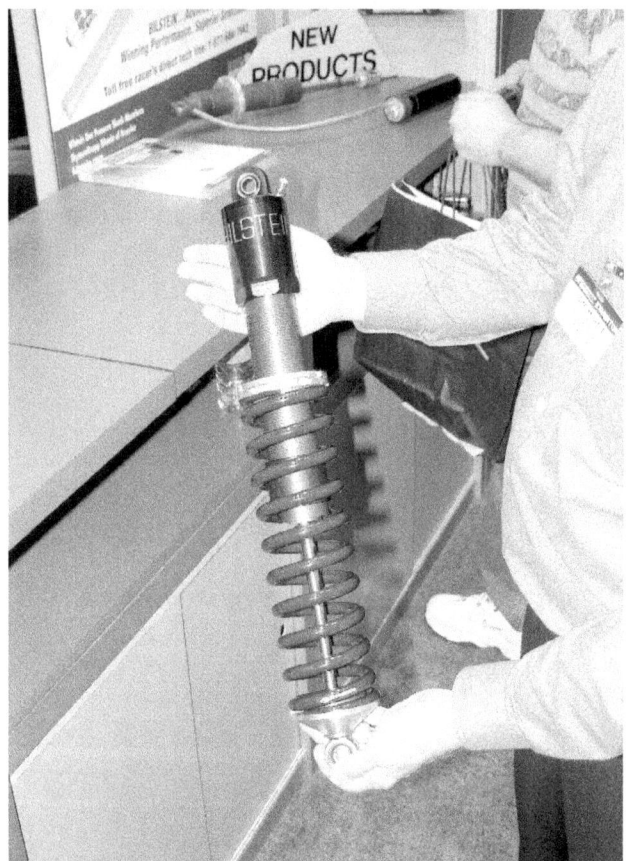

This is a modern coil-over shock and spring combination. We can see the black gas chamber at the top of the assembly. The shock body is threaded so that the spring nut atop the blue spring can be adjusted to preload the spring and control the ride height of the chassis and force on the tire.

Older non-adjustable shocks were built to specific resistances and numbered as to the stiffness. In this graph we can see the approximate force values for each number of shock. A one (1) shock was the least stiff shock. The left side shows the force in pounds and the bottom scale shows the speed of the shock movement in inches per second. As speed increases, the force also increases. This example has a digressive shock piston installed. It digresses in force after about 3.0 inches per second.

Low Speed Control – Low speed shock movement is defined as shaft speeds that are between 1 to 5 inches of movement per second. Many racers use 3 inches per second to evaluate the shocks resistance to normal entry and exit movements. The lower speeds are mostly associated with suspension movement caused by chassis roll and chassis dive at turn entry where the loss of speed is moderate. The low speed control dictates much of the handling side of the shock design.

Each shock has a piston mounted on the end of the shaft and one or more small "bleed" holes in the piston to allow fluid inside of the shock to flow freely from one side of the piston to the other. The size of the bleed hole regulates how quickly the fluid will flow back and forth and that is primarily how the different levels of resistance are created for low speed control.

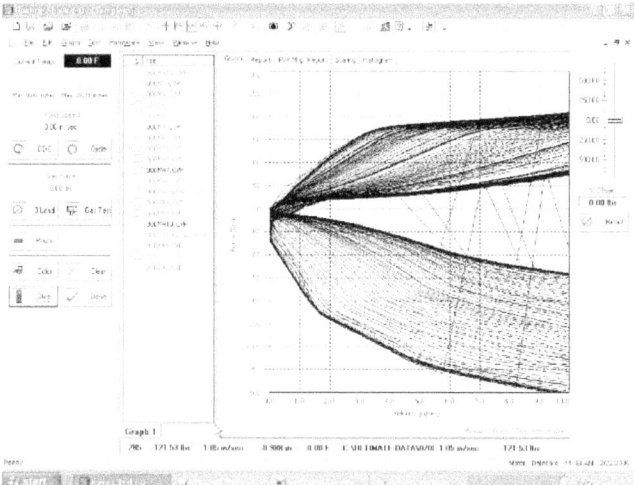

In a modern day adjustable shock, we can create almost infinite numbers of force settings. Race teams use these shocks to fine tune the transitions of corner entry and corner exit. Remember that the shock has control only when it is moving.

High Speed Control – As we experience the greater velocities of shaft movement we go into what is called high speed control with shaft velocities of from 5 to 10 or more inches of movement per second. Types of suspension movement that cause the higher shaft speeds in our shocks are: 1) bumps or holes in the racing surface (creating very high shaft speeds), 2) the driver stabbing the brakes on entry and hard on the throttle on exit, or 3) a sudden change in banking angle such as transitioning from banking onto the apron of the race track.

The piston mounted to the end of the shaft also contains a valving mechanism that allows the fluid to flow through slots that are designed into the piston. These valves consist of disks that open as the pressure increases due to more rapid shaft movement in either compression or rebound. These disks are used to control the dampening rate associated with higher shaft speeds.

Linear and Digressive Designs – There are two basic piston designs used in most race cars. One is the Linear piston that has a high flow rate at low shaft speeds and hence little resistance, and increases resistance as the shaft speed

increases. The rate of the shock continues to increase as long as the speed increases.

The other design is called a Digressive piston and has a lower flow rate at low shaft speeds which provides more resistance and control, and then the resistance rate of gain decreases with an increase in shaft speed to a pre-designed level. Then the resistance tapers off and stays relatively the same as the shaft speed continues to increase.

This "pop-off" characteristic works well to keep the shock from building excessively high amounts of resistance usually associated with sharp increases in shaft speeds.

A rut or hole in the track could cause the shock to move at rates well above 10 inches per second up to 20 to 30 inches per second or more. If that happened with a linear valved shock, it would feel like the suspension were solid with no suspension spring and the bump would jar the car considerably causing damage or loss of control.

Special Design Notation - If we study the automotive design books related to shock design for production automobiles, we would learn that the criteria calls for un-equal resistance in each direction, compression and rebound, in combination with the action of the springs.

Because springs naturally resist compression and promote rebound, we should not use a true 50/50 rated shock where the resistance is the same for both rebound and compression.

As the spring rate increases, the compression rate can be less and the rebound rate must increase. With bump devices with associated high spring rates, the rebound resistance must also be high in order to control the high spring rate.

Exam - In The Context Of This Lesson:

The Basic Primary Function Of A Shock Is?

1) Provide spring rate
2) Control motion of the suspension
3) Resist compression of the spring
4) Resist rebound of the spring
5) 2,3,4

A Shock Controls Motion By?

1) Adding spring rate to the suspension
2) Moving weight around on the four tires
3) Regulating fluid flow through the shock piston
4) Being pressurized

Compression and Rebound Are?

1) Equal in a perfect shock
2) Terms used to explain the motions of a shock
3) Made the same as the spring rate
4) The two types of pistons used in shocks

Compression Rates Should Be?

1) Equal to rebound in a perfect shock
2) Less than the spring rate
3) More than the spring rate
4) Intended to push down on the tire

Rebound Rates Should Be?

1) Equal to rebound in a perfect shock
2) Less than the spring rate
3) More than the spring rate
4) Intended to push down on the tire

High Speed Shock Control Is?

1) To control higher speeds of the car
2) The measure of control at faster shaft speeds
3) Less than that of low speed control
4) More than that of low speed control
5) 2 and 4

A Linear Shock Valve Does?

1) Keeps the shock shaft inline
2) Causes the force to be less at higher shaft speeds
3) Increases force with increases in shaft speed
4) Lines up the spring and shock on a coil-over

A Digressive Shock Valve Does?

1) Provides more control at higher shaft speeds
2) Causes the force to be more at higher shaft speeds
3) Diminishes the force with increases in shaft speed
4) Keeps the suspension from bottoming out

Lesson Twenty Five – Different Types of Shocks

Glossary:

Twin Tube Shock – A shock design that uses two tubes, one to hold a gas bag to separate the shock oil from the gas, and one to hold the shock piston, oil and shaft.

Gas Pressure Shock – A shock that uses pressurization of the gas to separate the oil from the gas.

Shock Body - The tube that holds the fluids and other parts including the shaft, separator piston and valve attached to the end of the shaft. The end of the shock body is attached to the chassis or control arm, depending on which way the shock is mounted.

Separator Piston – A free floating piston that is used to separate the shock oil from the pressurized gas.

Shock Piston – A part that is attached to the end of the piston shaft and to which is attached valve discs that regulate the flow of shock oil as the shock moves in compression and rebound.

Shock Valve – A series of discs of different thicknesses and diameters that are mounted on the shock piston that control the amount of fluid that passes through it when the shock moves in compression or rebound.

Bleed – A small hole, or needle valve, that allows the free passage of oil from one side of the shock piston to the other.

Shock Speed – The speed of movement of the shock shaft measured in inches per second. Generally speaking, as the shaft speed increases, so does the shocks resistance to movement.

Over View – All shocks are not of the same design. All shocks have a primary function to control the speed of suspension movement by regulating the flow of fluid that must flow through a valve system as the shock moves in compression and rebound. Simple shocks utilize simple methods to accomplish the required task.

All shocks must contain a volume of compressible gas to allow for the space taken up by the shock shaft as it is pushed into the shock body when the shock is in compression. As the shaft is pushed into the body, the gas compresses whereas the shock fluid cannot compress.

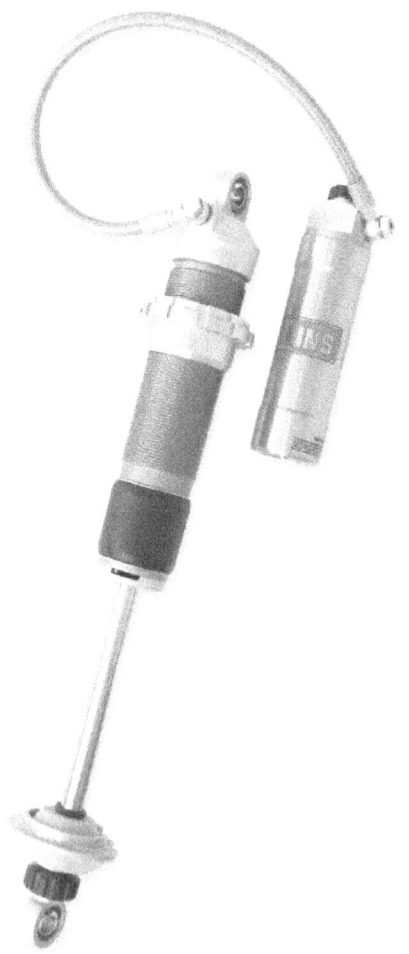

The most common shock used in today's racing is the gas pressure, fully adjustable shock. This one has an external gas canister that has its own valves to control compression settings. This type of shock can be different lengths and configured with different spring packages, including bump stops and bump springs. The functions remain the same though.

It has been shown and proven that high speed and repeated motion of the shock will cause mixing of the gas within the shock with the fluid if it is not made to be separate. As stated above, we need a volume of gas inside the shock body to allow the volume of the shock shaft to displace space within the shock.

If we don't separate the gas from the fluid, then they will mix and this affects the efficiency of the fluid passing through the shock piston and valves. The shock becomes less reliable as to the amount of force it provides. When the fluid is mixed with the gas, the combination flows more easily and therefore the force needed to move the shock the same speed becomes less and we lose efficiency.

This old style mono-tube shock seen here on a stock application is a rare sight in todays racing. "Stock" type shocks are now made for passenger cars that are of the gas pressure design, just like race shocks, and do a much better job of controlling the springs.

Shock Consistency – One of the main properties we require of our racing shock is that it stays consistent in how it provides resistance to the movement of the suspension. When the gas mixes with the fluid, the density of the mixture becomes less and then the resistance to the force also becomes less and not consistent.

Pressurizing the gas forces it out of the fluid and keeps the two substances separated. The gas is also physically separated by a floating piston in the design of a gas pressure shock, or by use of a gas bag in a twin tube shock.

The twin tube shock, although not formally pressurized, will have gas pressure as the shock shaft is pushed into the shock body and takes up space. When this happens, the gas in the bag must compress and therefore we have some amount of gas pressure in a twin tube shock.

Stock replacement shocks that are gas pressure are made by several shock companies like Bilstein and QA1. These do a very good job of controlling movement and staying consistent.

Differences Between Gas Pressure Shocks and Twin-tube – There is a distinct difference in the design and construction between the gas pressure shock and the twin-tube design. The gas pressure shock has only one tube to hold the fluid, the shock shaft and piston, the pressurized gas and the separator piston.

The twin tube shock has an inner tube that holds the fluid, the shock shaft and the piston, while another tube that surrounds the inner tube holds more fluid plus a gas bag. The fluid flows freely between the two tubes and the gas bag compresses as the shock shaft is pushed into the inner tube.

Primary Shocks Used For Racing – Now that we know the two primary types of shock design, the most commonly used shock design for racing is the gas pressure shock. So, we will concentrate on how that shock is designed for different applications. Much of what we discuss about the gas pressure shock will relate to the twin tube shock as well.

In the category of gas pressure shocks (GPS), there are different applications based on the different types of racing. This is directly related to what the shock is asked to do in each application. We can easily understand that the shocks will move much slower when installed in a race car that runs on a smooth, asphalt racing surface as opposed to a rough and rutted dirt track.

On asphalt, we can expect slower shock shaft movement in the range of 3 to 10 inches per second. These are the speeds where we design our asphalt shock rates to operate within. On dirt tracks, we can often see a range of shock shaft movement of from 10 to 25 inches per second or more.

The reason why the dirt shocks move so quickly is because the wheel moves that quickly when it hits a bump or hole in the track that may be inches deep. So, the shock built for dirt applications must necessarily be built differently.

Types Of Shock Piston - Within the description of different types of shocks we have different designs of shock pistons. The two basic designs are Digressive and Linear. Let's discuss the Linear piston first.

A linear design of shock piston starts out at a very slow speeds with very low resistance to movement. As the shock shaft speed increases, the rate of resistance increases in a linear way. In other words, if the speed doubled, then the rate also doubled. The rate doesn't have to double, but as the speed increases, so does the rate.

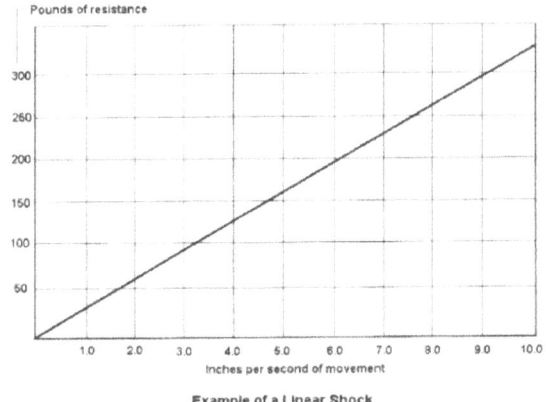

Example of a Linear Shock

One of the two main types of shocks and shock valving is the linear design. We can see the nearly straight line the force verses speed charted makes. As the speed of the shock shaft increases, the rate also increases proportionately. These shocks can achieve a very high rate of resistance if made to move very quickly. It can virtually feel like the shock bleed shut off and then the suspension becomes very stiff. It is best to only use this type of shock on a smooth race track.

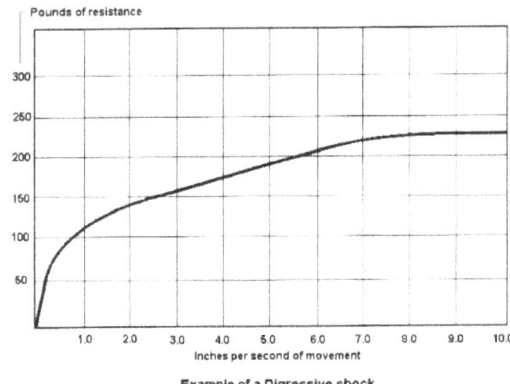

Example of a Digressive shock

The other type of shock and piston design is the digressive shock. For this design, as the shaft speed increases, the rate of the shock in pounds of resistance does not build proportionately. It will tend to taper off as the speed reaches a certain predetermined point. This allows the shock and wheel to ride over bumps that produce very high shaft speeds without producing high resistance. This design is much better for dirt applications where the tire rides over bumps and holes.

A digressive design of shock piston, when compared to a linear piston, has a higher rate of resistance to movement at slower shock speeds of say 1 to 3 inches per second. Somewhere around 5 inches per second (ips) the gain in rate begins to taper off and beyond 6 to 8 inches per second the gain in rate is minimal. Of course, we can design any rate curve we want, but generally speaking, this is the way a digressive shock piston works.

This is important to understand, because when we think about designing shocks for a dirt application, we have to think about those very high shock shaft speeds. Our shocks cannot keep building rate up to 25 ips or more shaft speeds or the shock would lock up. The fluids cannot move that fast from one side of the shock piston to the other.

If we use the digressive design of shock piston, then the rate would stabilize at some point and not gain and there would be a more free flow of fluid to allow the wheel to move over the bump. Otherwise, the suspension would lock up and the car would leave the racing surface and take off losing all grip.

Other Design Parts – The other designs available to make a shock what we want it to be for our application are many. The discs we stack onto our piston can be any number, and size and any thickness. The more disc and the bigger the size and thickness, the more they will resist flexing and the less fluid that is allowed to pass through the piston. The rate goes up.

Besides the valves on the piston that allow fluid to pass from one side of the piston to the other, we have what we call Bleed. Whereas the valves will be closed until they are forced open by movement of the shock shaft, the bleed will always allow fluid to flow freely from one side to the other.

At slower shaft speeds, there may not be enough force to open the valves whereas the bleed will always allow the flow of fluid, even at very slow speeds. So, the bleed controls the slow shaft speeds and the valves control the faster shaft speeds.

The use of shocks on race cars takes many shapes and forms. Even on this prototype rear suspension, the coil-over shocks with external canisters do the same job other shocks do in other applications. The only difference being the amount of travel and speed of movement between different applications.

On many adjustable shocks, the bleed is a needle valve located up near the shock piston. A rod is placed inside the shock shaft and extends to the heim joint where the shock connects to the control arm, or links thereto. We can see in the slot above the eye of the heim the bottom of the rod. This end has indentions where the team will insert a pin to rotate the rod and regulate the opening of the needle valve to adjust the low speed bleed.

The designs of bleed are varied. The bleed can be a simple hole drilled through the piston and the size of the hole determines the slow speed shock rate of resistance. Since the hole is permanent, it cannot be adjusted easily. You can drill a bigger hole, but you cannot make an existing hole smaller.

So, shock manufacturers decided to introduce a new way to bleed fluid. They make a hollow shaft that allows fluid to flow through it from one side of the piston to the other and the flow can be changed by adjusting the height of a needle valve within the shaft.

A simple screw adjustment moves the needle valve closer or farther away from a hole to make the space around it larger or smaller. For asphalt applications, obviously this type of adjustment works very well because we are working within a range of shock shaft speed that is slower and within the working range of the bleed.

For dirt applications, the bleed becomes less useful and we would need to work more so with the valve design to adjust for the higher speeds we see on rough dirt tracks.

Shock Graphs – The way we know how much work a shock is doing is by running it on a shock dynamometer. This device or machine measures the force of resistance of the shock at different speeds of shaft movement. It will record the forces in pounds or kilos and then produce a graph that can be printed out.

We would then look at the graph and note the force numbers for different shaft speeds. We want to match our shocks to the spring or wheel rates we have installed in the race car. Remember that shocks are primarily used to resist the spring wanting to rebound and the springs help the shock to resist compression. In almost every application, the shock will be deigned with more rebound rate than compression rate for the same shaft speed.

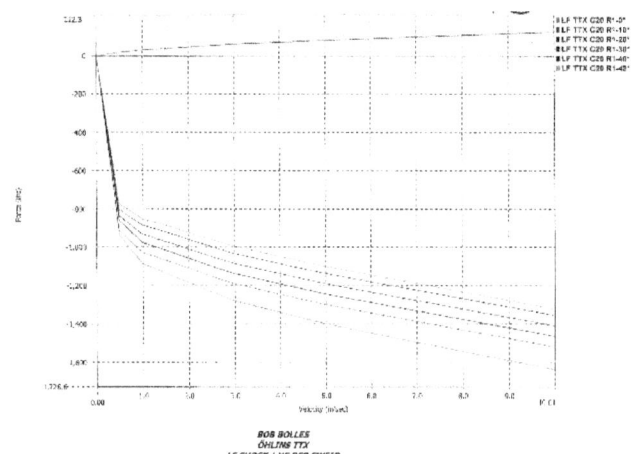

The use of a shock dynamometer and the printouts they provide are essential tools racers use to fine tune the setups of their cars in today's racing. In order to know exactly how much control a shock has verses the spring rate the shock will have to control, we need to look at the rate verses the speed of the shock in relation to what that corner of the car does on the race track.

So, to review, we have two primary types of shocks, the twin tube and the gas pressure. Within those two types, we have different internal designs of pistons, valves and bleeds. The type of piston, linear or digressive, and the size and number of valve discs and bleeds are chosen to control movement for the springs we have installed in our race car and the track conditions we will encounter.

And the shock dyno will tell us exactly what rates our shocks have for each shaft speed by producing a dyno graph we can read. Asphalt race teams will be more interested in the rates in the lower shaft speed range and dirt race teams will be looking at the rates in the higher shaft speed ranges.

Exam - In The Context Of This Lesson:

The Two Primary Types of Shock Are?

1) Twin tube
2) Gas pressure
3) Dirt applications
4) Asphalt applications
5) 1 and 2

What Is The Difference Between A Gas Pressure Shock And A Twin Tube?

1) The gps has a higher rate of resistance
2) The twin tube is better for dirt applications
3) A Gps has a single tube
4) The Twin tube shock has a gas bag
5) 3 and 4

The Following Is True Of Linear and Digressive Piston Designs?

1) The Linear is better for smooth tracks
2) The digressive piston allows faster shaft movements
3) One is good for dirt and the other good for asphalt
4) Digressive pistons have higher slow speed control
5) All of the above

Which Best Controls Slow Speed Shaft Movement?

1) The bleed design
2) The number, size and thickness of the piston valve discs
3) A linear piston
4) A digressive piston

Shock Graphs From A Shock Dyno Tell Us?

1) How much low speed control we have
2) How much high speed control we have
3) Resistance along a range of shaft speeds
4) If our shocks are good for dirt or asphalt

Lesson Twenty-Six – How To Control A Spring

Glossary:

Shock Piston – A part that is attached to the end of the piston shaft and to which is attached valve discs that regulate the flow of shock oil as the shock moves in compression and rebound.

Shock Valve – A series of discs of different thicknesses and diameters that are mounted on the shock piston that control the amount of fluid that passes through it when the shock moves in compression or rebound.

Bleed – A small hole, or needle valve, that allows the free passage of oil from one side of the shock piston to the other.

Shock Speed – The speed of movement of the shock shaft measured in inches per second. Generally speaking, as the shaft speed increases, so does the shocks resistance to movement.

Over View – The very reason we need shocks on a vehicle is to control the springs. Springs resist compression, or pushing them together. Once they are compressed, they want to expand to their free height, or the length they are when they are not loaded and compressed.

In a race car, or any car for that matter, the springs resist compression of the suspension and promote rebound of the suspension. If we didn't have shocks, the suspension would oscillate up and down freely. So, it has been known for a long time that we need to control the springs against this oscillation tendency.

For race cars, the primary function of shocks is to control the spring rate we have installed in the car. But there are other functions beyond that simple task that are useful. More on that later.

With racing shocks, we are trying to control the installed spring. In this design, the shock is mounted at a different motion ratio than the spring. Where it is mounted, it needs to control the spring as if that spring were converted to a coil-over spring, or much less than its actual rate. If we were to install a coil-over spring to replace the stock mounted spring, to maintain the wheel rate, we would need to use a much lower rate of spring. So, the shock mounted where a coil-over spring would be mounted needs to control an equivalent spring rate.

What We Are Trying To Control - When we refer to controlling the spring rate, we are mostly talking about the rebound of the spring. Since the spring resists compression and promotes rebound, the rebound motion is what the shocks primary job is to control. We don't need much compression control since the spring is helping to resist that motion and in fact doing most of the work.

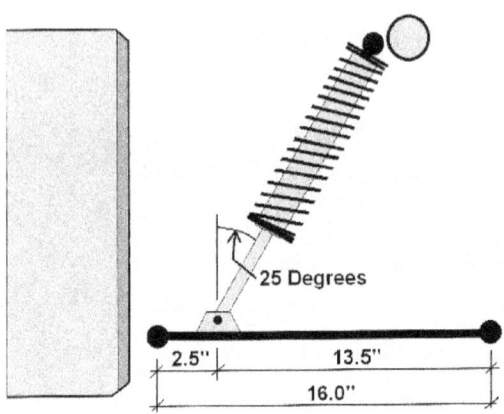

The shock that is a part of the coil-over system needs to control the spring on a one-to-one basis since it is mounted parallel to the spring and moves at the same speed and distance as the spring. This is much easier, unless we incorporate a bump, or dual spring system into the equation.

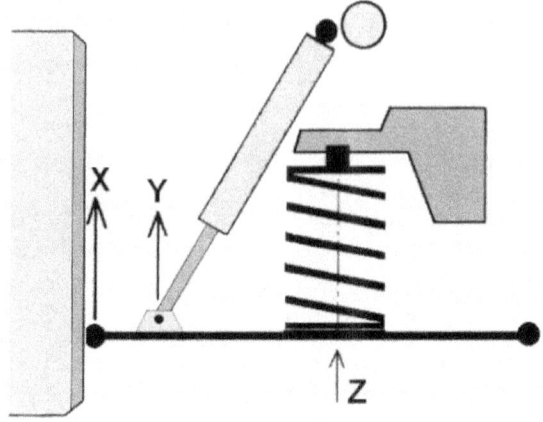

As stated in the first photo description, the shock in a stock system moves at a quicker speed than the installed spring and therefore can be setup to a lesser rate in order to control the stiffer spring moving slower. As the wheel moves at X pace, the Y end of the shock will move less and the Z center of the spring will move even less. The angle of the shock to the motion of Y reduces the shock speed even more.

When we talk about controlling the spring, we are referring to controlling the specific spring rate. Naturally the higher the rate of the spring, such when going from a 100 pound per inch rate to a 200 pound per inch rate, we would need to double our shock rate of resistance to rebound.

There is no set or average shock rates, only what will control a specific spring rate. So, if we install very stiff springs such as with road racing cars or formula type race cars, the shocks must have sufficient rebound rate to control the very stiff spring. In other words, the shock rebound rate must also be very high.

Compression Control - Conversely, if the spring rate is high, then that spring will resist compression at a high rate too. Our shock compression rate would be very small because the spring is doing all of the work resisting the compression motion.

Another example of high spring rate in a race car is the current trend of stock cars running on bumps. Bumps are devices that provide another spring that is higher than the ride spring, and in many cases, much higher. The bump can be either a bump stop using very stiff material or bump springs which are coil springs rated very high in pounds or kilos per inch.

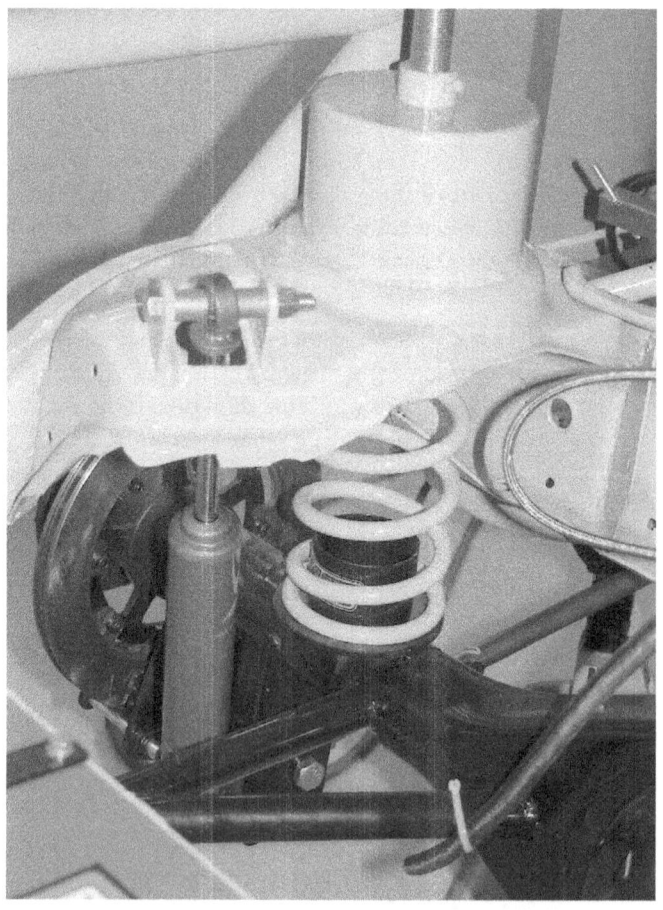

In this design, the shock is moving as if it were a coil-over shock, inline with the spring and at the same speed and distance in chassis movement. So, the team would look at the proper speed of movement for that type of track and then choose a shock rate of rebound resistance that matched the spring rate.

The shocks these teams use in conjunction with the bump setups have to control the high spring rate of the bump and so they are built to have a lot of rebound resistance. These shocks are wrongly referred to as tie-down shocks because they overcome the much softer ride spring and keep the chassis down on the bump device. In reality, even though they do hold the chassis down, their real function is to control that very stiff bump device and its tendency to rebound more aggressively.

So, we control a spring by using a shock that is rated in rebound resistance equal to the rate of the spring. In compression, the spring is doing most, if not all, of the work and therefore we need much less compression control in our race shocks.

This is a stacked spring. It works like this. The two springs when stacked and working free are a lower rate than either of the springs individually. The top spring has a higher rate than the bottom spring. When the shock has compressed a pre-determined amount as the car goes through the turns, the divider ring hits a stop and only the top spring is then in use. Since it is a higher rate than the combined rate of the two springs together, the overall spring rate on that corner of the car goes up. And the shock must control the higher rate.

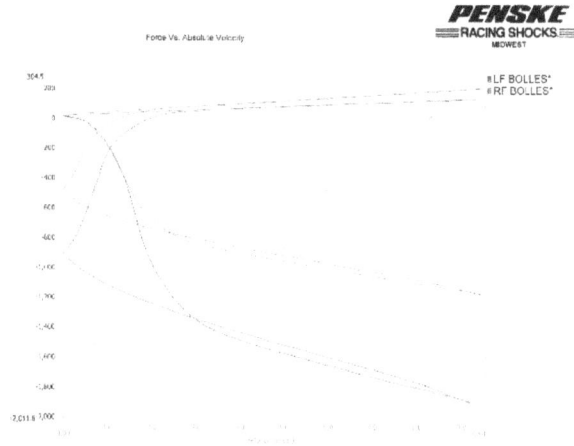

For bump setups, or stacked springs, as the spring rate spikes, so does the shock rate in rebound. The lines below the zero (0) represent the rebound settings and rate. This shock has a rate of resistance for the Right Front (RF) shock of nearly 500 pounds before the shock even moves. This is called "nose". This keeps the shock inside the spring rate range of the bump. Note that the left front shock has a nose rate of 900 pounds to control a heavier bump device. Note also that the compression rate, the lines above the zero on the left scale are next to nothing. The heavy bumps resist compression, so the shock doesn't have to do any work.

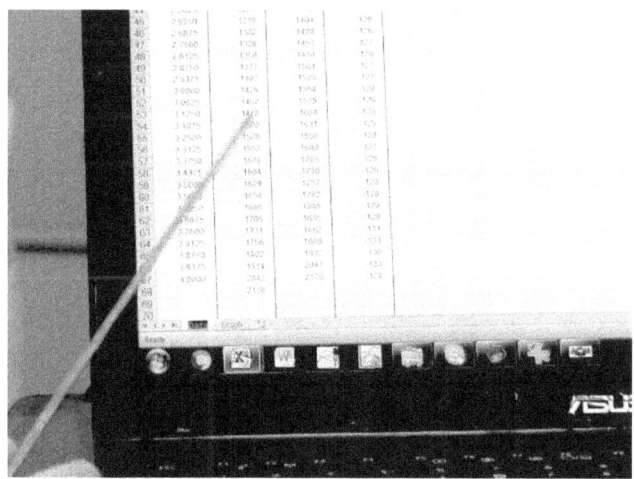

The Gale Force Load Pro measures the force of the spring as it travels through a range equal to what it does on the track. It records and shows the rate per each 1/16th of an inch throughout the travel range. In this way, the user can see exactly when the rate begins to rise and when it spikes as the bump nears maximum compression.

Exam - In The Context Of This Lesson:

The Primary Reason We Need Shocks Is?

1) Slow the motion of the chassis

2) Provide loading at one or more corners

3) To control the installed spring rate

4) To provide compression resistance

What Needs More Control?

1) The front end on braking

2) The rear on acceleration

3) Rebound of the springs

4) Compression of the springs

Running High Spring Rates Or Bump Setups Require?

1) Less rebound shock control

2) More compression shock control

3) Less compression shock control

4) More rebound shock control

5) 3 and 4

Lesson Twenty-Seven – Motion Ratio Affects Shock Rate

Glossary:

Lower Control Arm – A part of a Double A-arm suspension. In most cases, the springs are mounted to the lower control arms.

Coil-Over Shock – A shock where the spring is mounted over the shock body. The spring moves the same distance and at the same speed as the shock when the wheel moves.

Wheel Rate – The installed spring rate translated out to the wheel. The wheel becomes a spring rate as if it were an actual spring and it is always less than the actual installed spring rate.

Motion Ratio – A ratio that defines the difference in movement between two parts in a suspension. Common motion ratios of interest in a race car are: wheel to spring, and spring to shock ratios.

Over View – We now know how a shock controls the spring rate in the race car. We know the shock resists rebound and the spring mostly resists compression. But what if the spring and the shock are moving at different speeds? How do we compensate for the difference in speed of movement? That is what you are about to learn.

Although this spring mount has many motion ratios through the rockers, etc. the spring and shock are still in a one to one ratio. The movement of the shock/spring divided by the movement of the wheel equals the wheel rate motion ratio. On this type of car the shock could move less or more depending on the design, but the shock to spring ratio is still one to one.

In almost every race car, the wheel moves at a different speed than the spring and shock. If we want to control the speed of movement of the wheel, then we need to know what to do with our shock rates to accomplish that.

In this example, we have a shock that moves at nearly the wheel speed and a spring mounted well inboard of the wheel that moves much slower. So, the shock will be more efficient than if it were mounted to where it moved at the same speed as the spring. If it took a certain rate to control this spring on a one to one ratio of movement, then with the shock moving faster, and having a greater rate due to that speed, we can use a shock rated at a lower value to do the same work of controlling the spring.

Motion Ratio Explained – A motion ratio is when one part moves more or less than another connected part. In the case of springs and shocks, the wheel always moves a greater amount than the spring or shock. And the shock may move a different amount than the spring. If the spring were mounted on the shock, then the spring to shock ratio would be one to one. But the spring and shock are almost never mounted to where there is a one to one ratio to the movement of the wheel with one exception.

On some prototype and formula cars, there are rockers and pushrods that transfer the spring rate to the wheel. The ratio of the rocker could conceivably cause the spring and/or shock to move farther and faster than the wheel.

So, we can assume that for every inch the wheel moves, the shock and spring are moving either more or less. For stock cars, the closer the spring or shock is mounted to the chassis mount and away from the wheel (ball joint), the less it moves in relation to the wheel.

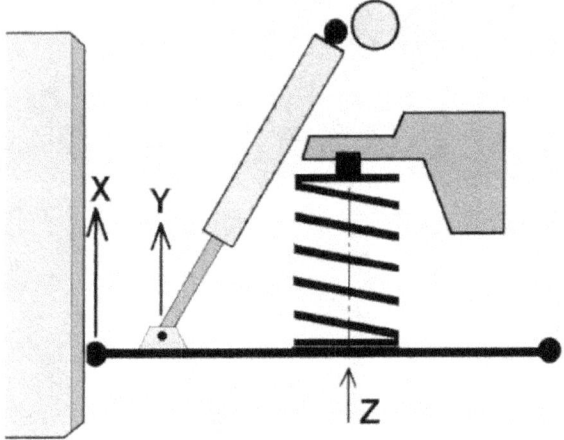

Here we see the actual condition described in the last photo. The wheel moves at X speed, the shock moves at Y speed, a little slower than the wheel, and the spring moves at Z speed, much slower than the wheel or shock. If we could find the motion ratio between the movement of the shock and that of the spring, we could multiply that number times the shock rating to find the rate needed to control the spring with this motion ratio. We will do exercises where we determine the exact shock rate to use in RCT Level Two.

For a shock that is mounted half way between the ball joint and the inner control arm mounts, the motion ratio for wheel to shock would be 2:1, or for every two inches the wheel moved, the shock would move one inch. Inversely, the shock motion ratio to the wheel would be 1:2, or one inch of shock movement to two inches of wheel movement.

This is an important consideration when we are designing our shocks to control wheel movement. And, if the shock is mounted at a different motion ratio than the spring, that again is a consideration when designing the shock to control the spring rate.

Which Direction To Go – For race cars running coil-over shocks where the spring is mounted on the shock body and moves at the same speed as the shock, we basically need to control the spring with the same shock rate as the spring rate for a certain predetermined shock speed. There can be variations, but the statement is basically true.

This is a very good example of how a spring moves at a different speed than the shock. On this car we call a swing arm, or Z-link, the spring is mounted onto the lower control arm. The device holding the spring is a slider to locate the spring and it is not a shock. The shock is mounted onto the rear axle tube on more of a one to one ratio of movement to the wheel. In this case as in the last example, the shock moves quicker than the spring and therefore needs less rate to control the spring than if it were a coil-over design.

If the shock were mounted away from the spring, then we need to study and know the motion ratio of the spring to shock movement. This design is common on stock cars where a large spring supports the car and is mounted to the lower control arm and the shock is mounted away from the spring and usually closer to the ball joint.

In this case, the shock would move a greater distance and at a higher speed than the spring. Since shocks gain rate with higher speeds, we would necessarily need to reduce the rate of the shock from a what we would require for a one to one ratio to something less.

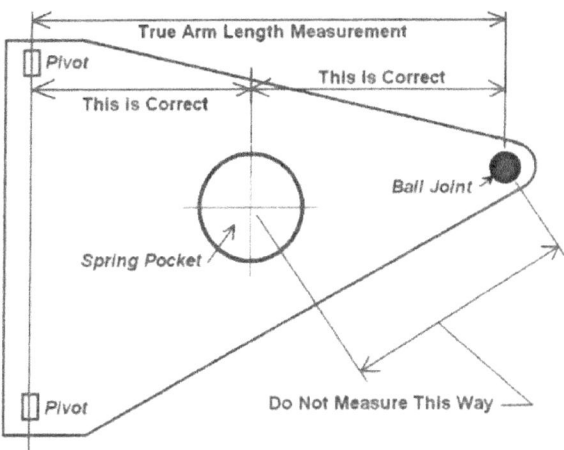

When measuring for motion ratio calculations, be sure to measure correctly. We want the length of the control arm to be the rotational radius of the arm. So, we measure the distance from the pivot line to the center of rotation of the ball joint at a ninety degree angle, as shown.

In some prototype cars, it is now customary to mount the shocks in a different location than the springs. Again, we must know if the shock is moving faster or slower than the spring in order to rate it to properly control the spring rate.

If the spring were moving faster, then the shock would need to be rated higher than the spring rate. If the spring were moving slower, the opposite would be true, we would reduce the rate of the shock.

We now know that motion ratios affect how the shock controls the spring and wheel movement. In designing a race car, we must know the motion ratios and how those affect the control of the spring rate.

Exam - In The Context Of This Lesson:

A Motion Ratio Is?

1) How far a wheel moves in the turns

2) The amount the wheel moves versus the spring

3) The ratio of shock to spring movement

4) 2 and 3

A Wheel To Spring Ratio Could Be?

1) One to one

2) 1:2

3) 2:1

4) Less than the shock movement

5) All of the above

If The Spring Is Moving Faster Than The Shock?

1) The rate of the shock must be less than the spring

2) The rate of the shock must be more than the spring

3) The motion ratio is 1:2

4) The motion ratio is 2:1

For Some Prototype Cars The Spring/Shock Moves Faster Than The Wheel Because?

1) The motion ratio of the spring to wheel is less than one to one

2) The rocker design motion ratio causes quicker spring/shock movement

3) The spring/shock is mounted away from the lower control arm

4) The push rod is mounted at an angle to the lower control arm

Race Car Technology – Level One
Lesson Twenty-Eight – Race Tire Basics

Glossary:

Tire Camber – The angle a tire makes from vertical relative to the racing surface. Positive camber is when the top of the tire is tilted to the outside from the chassis. Negative camber is when the top of the tire is tilted in towards the chassis.

Tire Temperatures – The readings of the core temperatures across the face of a race tire after it has been run a sufficient number of laps to reach operating temperatures.

Tire Contact Patch – The part of the tire that contacts the track surface. The area and loading of the contact patch is the most important consideration for race teams and everything else we do in race car setup and design is intended to maximize the tire contact patch size and loading.

Over View – This entire Course is about race car technology, defining the parts that make up a race car and how those parts fit together. The ultimate goal for race car engineers and crew chiefs is to find the most Grip so that the car will go faster through the turns.

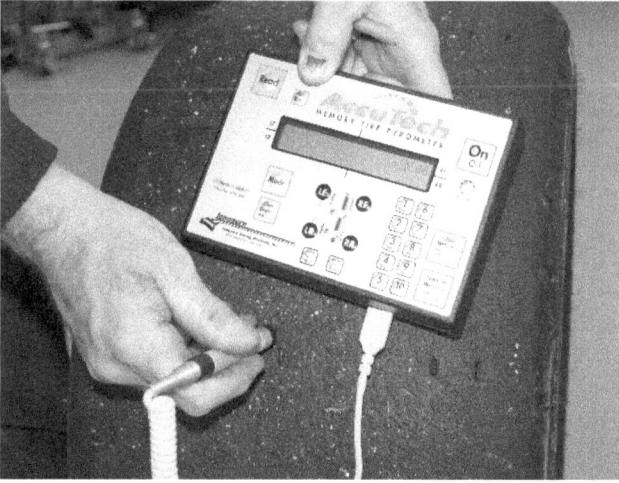

We evaluate how our tires are working on the race track by taking tire temperatures, as shown here, taking tire pressures and by evaluating the tire wear. We use this data to determine what settings to use for pressures, camber and setup that will produce the greatest performance.

In working with the race tires, we learn how to develop more grip for the whole car. Speed gained through lateral grip in the turns will be add to the average speed over the entire lap. That statement is probably the most important thing to understand and grasp of the entire course.

Tire Characteristics – Tires have more grip as they become more heavily loaded, or have more weight on them, and when the contract patch is made larger. So, we try to design our race car so that we can maximize the amount of loading on the four tires and provide the largest contact patches for those tires.

One important thing to understand about tires is this: Tires gain traction or grip when adding load in a non-linear way. If we double the amount of load on the tire, it gains less than double the grip. So, for a pair of tires on the same axle, or same end of the race car, the most grip we can obtain for that end of the car is when both tires are equally loaded.

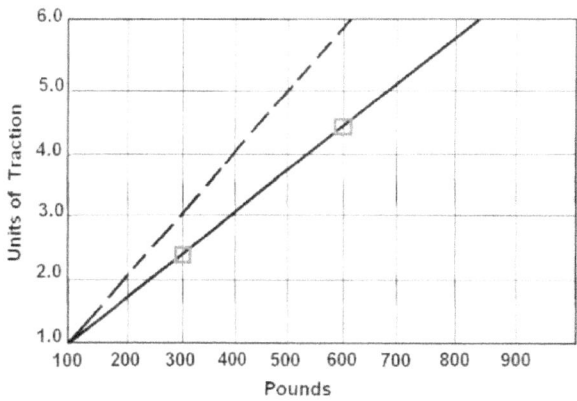

When we add load to a tire, the grip, or traction, the tire has does not increase at the same rate as the increase in load. The grip increases less. So, if we double the loading on the tire, the grip increases by less than double. This concept is one of the most important to understand and is the essence of understanding race car setup and balance.

For most race car applications, we cannot equally load the tires, so we try to get them to be the closest to equally loaded. With load transfer happening in the turns, the inside tires end up with less loading than the outside tires in almost every case. The more load we can retain on the inside tires, the more equally loaded the two tires will be and the more grip we will have.

As for contact patch, our tire pressures mostly dictate the optimum contact patch area that will distribute the

125

loading in the most efficient way. Then we will be able to utilize the loading we have produced by designing the best setup.

Tire Temperatures – Tire temperatures tell us a lot about how the tire is working and how much grip we can expect from our tire contact patches. We usually measure the temperatures of the tires as quickly as possible after the car has completed enough laps to properly heat the tires.

The proper way to take tire temperatures is to read the temps in three places across the face of the tire with a tire temperature probe. This instrument has a needle like end that is inserted into the rubber to get the inside temperature. It should be inserted at a 45 degree angle into the rubber and at the edges about one inch inside the very edge of the inside and outside of the tire and at the middle of the tire.

The proper way to insert a temperature probe is to push it in at a 45 degree angle. Take the outside temperatures in from the outside edge of the tire approximately 1 to 1 ½ inches. Allow enough time for the reading to settle before recording and moving on. Then take a middle of the tire reading, and then an inside of the tire reading again about 1 to 1 ½ inches in from the edge.

Give the instrument time to react to the temperature of the tire. Once the reading stabilizes, record the amount and move on to the next reading. Don't be in too much of a hurry to do this because a lot of decisions will be made based on the temperature readings. They must be accurate.

Tire Pressures – For every type and use of a racing tire, there is an optimum pressure that will provide the best possible contact patch area and therefore the maximum amount of grip. Most race teams will use Nitrogen gas to inflate their race tires because it contains very little moisture. Moisture when heated expands and causes a greater increase in the tire pressures.

For taking tire pressures, always try to use a large dial gauge. Teams will sometimes set tire pressures in half pound increments and a larger dial makes this easier to do. Tire pressures should be taken a soon as the car is off the track. We can read the tire temperatures to tell if our tire pressures are correct.

When race tires are run at high speeds through turns, they heat up. The rubber heats the gas inside the tire and it expands. When it expands, the shape of the tire changes. Since our contact patch shape is so important, we want to make sure that when the pressures have increased that the final pressure will create the best contact patch shape. Excess gains in the tire pressures will cause a reduction in the contact patch size.

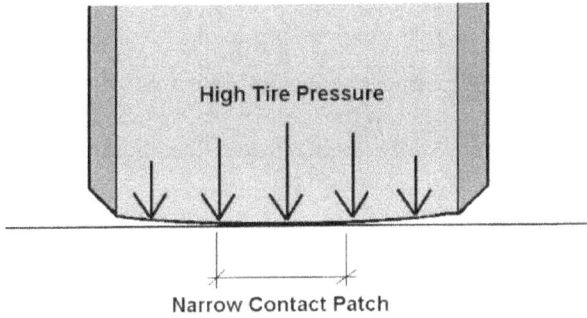

If the tire temperatures are too high at the middle of the tire, the tire pressures are too high meaning that the middle is working harder than the outsides of the tire. This causes a reduction in the size of the contact patch resulting in loss of grip. The middle temperature should be the average of the two outside temperatures.

We try to predict the pressure changes for each tire so that we can start out with lower pressures than optimum with the goal of ending up with the correct pressures and greatest area of contact patch. This takes time and experimentation because each car and track is different in the way the tires heat up and expand.

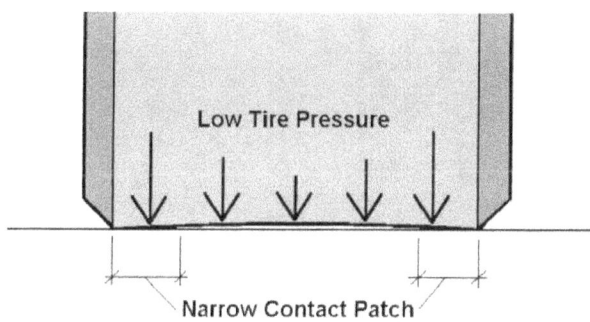

If the middle temperature is too low, the tire pressures are then too low. This also causes a loss of contact patch area and loss of grip. We need to add pressure until the middle temperature is an average of the outside tire temperatures. The problem is not that all of the width of the tire is not on the ground, it is the fact that whole area of the contact patch is not exerting the maximum pressure on the racing surface like it would be if the pressures were correct.

Running and Evaluating the Tires – When we run on a set of tires at race speeds, we need to look at three things to determine how well the tires are working. We look at the tire Temperatures, tire Pressures, and tire Wear.

Chart 1 — Session #3

Left Front

Loc 12	Loc 11	Loc 10
146	154	150
Avereage Temp. 150.00		
Pr. Cold	Pr. Hot	Pr. Gain
20	23	3

Right Front

Loc 1	Loc 2	Loc 3
211	204	207
Avereage Temp. 207.33		
Pr. Cold	Pr. Hot	Pr. Gain
24	29	5

Front Average = 178.67

Left Rear

Loc 9	Loc 8	Loc 7
162	168	158
Average Temp. 162.67		
Pr. Cold	Pr. Hot	Pr. Gain
20	25	5

Right Rear

Loc 4	Loc 5	Loc 6
216	214	210
Average Temp. 213.33		
Pr. Cold	Pr. Hot	Pr. Gain
24	32	8

Rear Average = 188.00

In this chart of tire temperatures, we can see several problems. At the left front, the Loc 11 tire temperature is higher than the two outside temperatures for this tire. We would need to lower the tire pressure. At the right front, the Loc 2 temperature is cooler than the average of the two outside tire temperatures and this tire would need more tire pressure. As far as setup, the left front tire average temperature of 150.0 deg. Is cooler than the left rear average temperature of 162.67. This indicates a setup that is unbalanced and needs correction. In RCT Level Two we will tell you how to do that.

The last two are fairly easy to read for both asphalt and dirt teams, but the first may be difficult for dirt racers to get due to the way heat bleeds off the tire after a run. None the less, if possible, get all three to help understand all that the tire is doing.

Chart 2 — Session #4

Left Front

Loc 12	Loc 11	Loc 10
148	152	157
Avereage Temp. 152.33		
Pr. Cold	Pr. Hot	Pr. Gain
18	22	4

Right Front

Loc 1	Loc 2	Loc 3
212	208	204
Average Temp. 208.00		
Pr. Cold	Pr. Hot	Pr. Gain
26	32	6

Front Average = 180.17

Left Rear

Loc 9	Loc 8	Loc 7
164	162	159
Average Temp. 161.67		
Pr. Cold	Pr. Hot	Pr. Gain
18	23	5

Right Rear

Loc 4	Loc 5	Loc 6
214	212	209
Average Temp. 211.67		
Pr. Cold	Pr. Hot	Pr. Gain
26	34	8

Rear Average = 186.67

In evaluating this tire temperature and pressure chart for a circle track race car turning left, we can see more problems. At the left front, the inside tire temperature at Loc 10 is hotter than the outside tire temperature. This is reverse of what we want to see for this type of car. The right front looks fairly good with the middle temperature about average of the outside temperatures and the inside temperature is higher than the outside temperature. Also, we see where the right rear tire pressure is higher than the right front meaning the car is a slight bit loose causing the right rear tire to be hotter.

The temperatures tell us how well the tire footprint is working. It would seem that the heat ideally should be close to even across the face of the tire. This all depends on the type of race car and where it is racing.

Chart Four
Session #6

Left Front				Right Front		
Loc 12	Loc 11	Loc 10		Loc 1	Loc 2	Loc 3
163	163	162		204	204	203
Average Temp. 162.66				Average Temp. 203.66		
Pr. Cold	Pr. Hot	Pr. Gain		Pr. Cold	Pr. Hot	Pr. Gain
17	21	4		25	32	7

Front Average = 183.16

Left Rear				Right Rear		
Loc 9	Loc 8	Loc 7		Loc 4	Loc 5	Loc 6
164	163	159		204	202	201
Average Temp. 162.00				Average Temp. 203.00		
Pr. Cold	Pr. Hot	Pr. Gain		Pr. Cold	Pr. Hot	Pr. Gain
17	21	4		25	32	7

Rear Average = 182.50

Reading this chart, we see what appears to be a perfect setup and tire cambers. The averages for each side are the same, the pressures for each side are the same, and the tire temperatures across the face of each tire are nearly the same. This used to be what we looked for, but more recent experimentation has led us to a different view of front tire temperatures that we will discuss here.

In modern stock cars racing on circle tracks, most teams prefer to see more heat on the inside (towards the inside of the race track) of the tire, but the progression of the heat should be steadily increasing from the outside to the inside in any event.

The higher heat on the inside part of the tire shows that it is more heavily loaded. When that part of the tire is more heavily loaded, it expands or flattens out. This creates a shape that produces a larger contact patch area than if the tire temperatures were even across the tire.

This scenario only works with high profile tires where the sidewalls are tall and fairly soft as opposed to low profile tires with stiff sidewalls. In road racing, where we see the later type of tires used, this practice would not apply.

How To Make Changes – The team should not make major changes to their tire pressures or front cambers based on the tire temps. until they have had a chance to run the car at full speed for more than five laps. We need to get maximum temperatures in the tires in order to see the true picture of what the tires are doing.

If the temps are lower or higher in the middle of the tire, it is an indication of lower or higher than adequate tire pressures. A low temperature at the middle means the pressure is too low and the middle of the tire is not pressing hard enough against the race track surface to create the ideal heat.

If the middle temperature is too high, that means the tire has too much pressure and the middle is working harder than the outside edges of the tire. We try to maintain an even spread of temperatures across the face of the tire, not equal temperatures necessarily.

The tire wear also tells us a lot about how a tire is working. If we measure the depth of the tire grooves or wear slots across the face of the tire, we can see if the cambers and pressures are correct. This is especially useful for situations where taking tire temperatures is not practical, such as when racing on dirt.

For some types of racing it is very difficult to take tire temperatures. So, we can take readings of the tire tread depth to get an indication of how the tire is working. More wear on one side of the tire verses the other tells us we need to make changes to the cambers. If the middle is wearing less than the outsides, then we need to increase the tire pressure to make the middle work harder and wear more as a result.

The depths can be read similar to how we read tire temperatures. If the middle of the tire is worn more so than the outsides, the pressures are probably too low. If the outside of the tire is wearing more than the inside, the cambers are probably too little and the top of the tire needs to be leaned in towards the inside of the turns.

Tire Cambers Vs Contact Patch Size – The final layout of the tire temperatures and the resulting tire contact patch shape and size are dependent on the tire camber and tire pressures. These two conditions work together to produce the contact patch loading for each tire.

Race teams are constantly experimenting by adjusting the camber angles and the tire pressures to find the best grip that will cause the greatest speed through the turns. The pressures cause the contact patch to be consistent in shape across the face of the tire. The camber angle optimizes the size of the contact patch.

Lesson Three tells us about cambers and camber change. This area of technology for race cars is one of the most important to understand in todays racing. More grip has been found by working with cambers and the geometry that maximizes cambers than with any other area of race car technology.

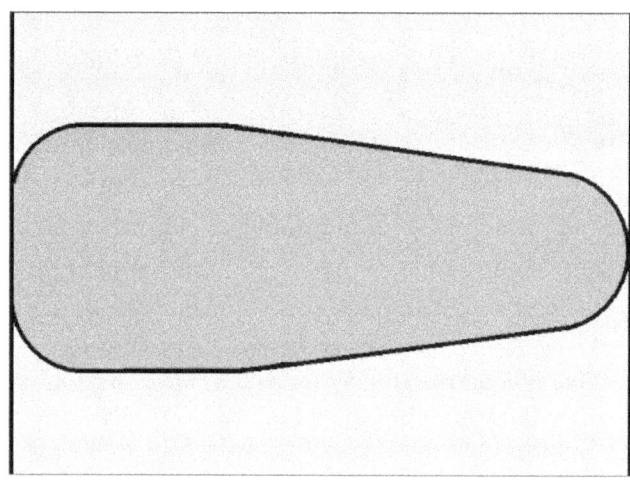

In the recent past for circle track race cars and formula cars with high profile tires and fairly compliant sidewalls, we liked to see the inside tire temperatures higher than the outside. This created a larger tire contact patch and helped the car to turn better. As teams experimented more with cambers, this process went a little farther.

When we record even tire temperatures across the face of the tire, this is what the tire contact patch looks like. For road racing applications, this may be as good as it gets for low profile tires with harder sidewalls. For circle track racing and for any application where the tires used are taller and more compliant, there might be a better design.

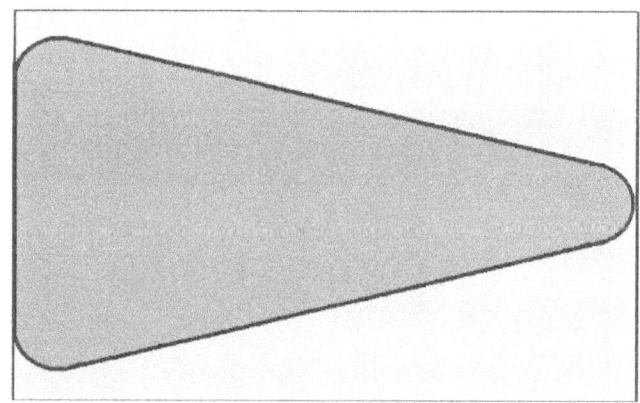

For tires with taller sidewalls that are more compliant, we can run much more camber than was thought possible in years past. This creates more loading and heat on the inside of the tire (towards the radius of the turn) which produces the contact patch shape shown here for a left turning car. This contact patch area is larger than what we would see if the tire temperatures were even across the face of the tire. It has a larger area and therefore creates more grip to help the car turn.

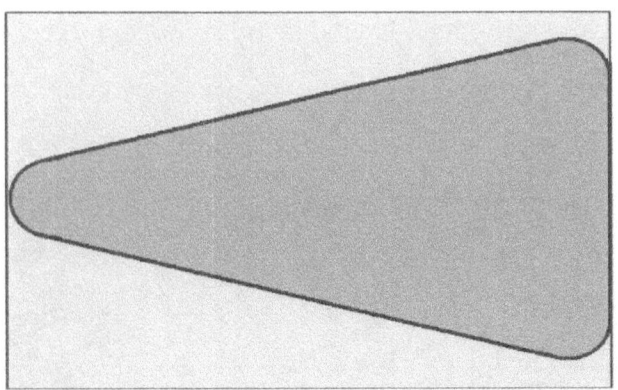

The rear tires on a straight axle suspension cannot be easily adjusted for camber angles. So, if there is no adjustment, how can we maximize the contact patch shape. If we allow a lower pressure to be run in a tire on a straight axle turning left, then we might end up with this contact patch shape. The temperatures must still be an even progression across the tire face with the middle being an average of the outside temperatures.

Storing Tires – Tires will lose their grip properties quickly with age. The chemicals that produce a softer tire compound evaporate starting right after they are manufactured. It is best to cover new tires with a wrapping and store them in a dark, cooler place to delay the aging process.

Used tires can always be run in practice or testing providing we understand that they will provide less performance than newer tires. If we know the difference, then we can accurately determine if we are competitive without needing to compare lap times directly.

Tires used for racing are softer than those used for road car applications. The compounds and chemicals that make this possible will evaporate over time. So, from the time a tire is produced until it is used, it degrades and becomes harder. A harder tire has less grip than one that is softer, to a point. A tire can be too soft and there is a point where maximum grip is reached and the tire will still hold up. These tires are wrapped and covered to prevent degradation and loss of grip properties.

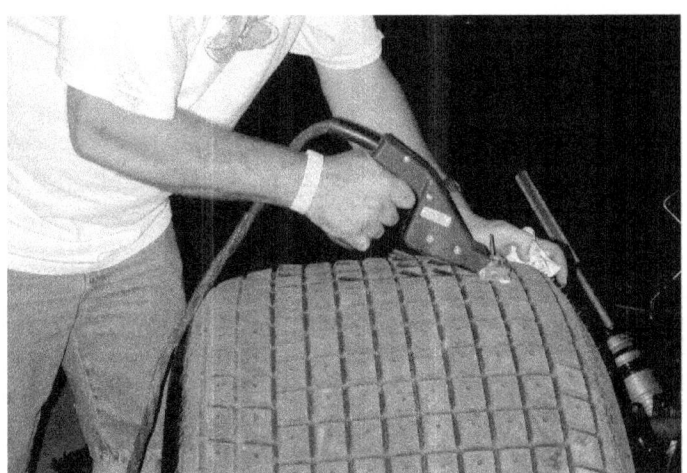

There is a lot more technology associated with racing tires and depending on the use and type of racing, more to teach about. In RCT Level Two and Three we will discuss in detail how each class of racing works with their tires.

Exam - In The Context Of This Lesson:

When We Double The Load On The Tire, We?

1) Double the grip

2) Create more than double the grip

3) Create less than double the grip

4) Can run stiffer springs

Tire Pressures Help The Tire How?

1) Develop the best tire contact patch

2) Keep the temperature of the tire down

3) Creates more tire temperature

4) Maintain equal loading across the face of the tire

Ideal Camber Settings Help The Tire How?

1) Bring equal temperatures across the face of the tire

2) Keep the temperatures lower

3) Create a larger contact patch

4) Keep the middle of the tire cooler

Tire Pressures Change Because?

1) The valve stem leaks

2) The track surface changes

3) The cambers are not correctly set

4) The tires get hot

Low Profile and Hard Sidewall Tires Need?

1) To have a hotter inside temperature

2) Lower tire pressures

3) Even tire temperatures across the face of the tire

4) Higher tire pressures

Tires With A Tall And Softer Sidewall Need?

1) To have a hotter inside temperature

2) Lower tire pressures

3) Even tire temperatures across the face of the tire

4) Higher tire pressures

The Best Time To Make Changes To Tire Settings is?

1) After the first run of the day

2) Right before the race

3) After getting the tires up to optimum temperatures

4) Once the air temperature stabilizes

The First Thing To Make Changes To Is?

1) Tire cambers

2) Cross weight

3) Toe settings

4) Tire pressures

Lesson Twenty Nine – Tire Stagger

Glossary:

Circumference – The distance, or measurement, around the outside of a tire to determine how far it would roll in one complete revolution.

Rear Differntial – A mechanical device that compensates for the difference in radius of the outside and inside tires when going through a turn and allows the two tires to turn at the correct speed.

Tire Stagger – A difference in circumference for two tires mounted on the same axle, or end, of the car. Stagger is a design function to allow for different radii of the inside and outside tires when going through a turn.

Spool – A mechanical device that locks the two axles in a rear end so that they turn at the same speed.

Over View – We learned early on in this course that the ultimate performance of a race car is dependent on how well the tires Grip the race track. We will now provide you with information on how we work with the tire stagger to gain the most Grip.

This information is primarily intended for race cars that only turn in one direction such as circle track race cars that run in the U.S. and other parts of the world. For road racing cars, we will most often select tires that are equal in circumference on each side at the same end of the car. So, stagger is not an issue with those cars.

Selecting tire sizes and predicting changes in stagger is the first step in working with race tires. For race cars that only turn one way, we utilize stagger in the rear end,, or whichever end drives the car, so that each tire will be rotating the correct speed to match the radius of the arc they will follow around the turn.

There are various rear differentials that race teams use to lock and un-lock the rear axles when entering (un-lock) and exiting (locked) the turns. When the axles are locked, the tire circumferences must be correct or one or both tires will slip and that end of the race car will lose traction.

On circle tracks, when going through the turns, the outside tires turn at a different speed than the inside tires if they are the same diameter. If the tires are connected to a solid axle system, they would fight each other at different speeds and cause handling problems. Because we want the tires to turn at the same speed, we need to introduce stagger, or a difference in diameter of the tires, to allow the tires to roll around their own radius easily. The outside tire would be larger than the inside tire.

Matching Stagger to the Race Track - One of the most important things to know is that rear stagger should match the race track. Using more or less stagger than the race track requires will cause loss of grip.

It was at one time common to try to fix handling problems by adjusting the stagger. This is not recommended and is considered a crutch, or false way to solve a race car handling problem.

The tire sizes cold and with the same pressures are marked on the tires in chalk for these circle track racing tires. Stagger is achieved by selecting different sizes and using different pressures for the left and right side tires.

Each circle track racetrack has a turn radius (average radius of both turns) that will require a certain rear stagger amount. That stagger number represents the difference in circumference of the two tires on the same axle, the left being the smaller one (for left turns), that would result in both tires turning the same number of revolutions through the middle of turns.

Type of Rear Differential - Stagger influences a car that is using a spool rear differential (both axles locked together at all times) on entry to the turns as well as through the middle and off the corners. With a "Detroit Locker" type of rear differential, one of the rear wheels is unlocked when entering the turns. The Detroit Locker will then lock both rear wheels under acceleration and stagger mostly affects handling from mid-turn and off the corners.

Less Stagger - Too little rear stagger will try to turn the car towards the outside wall and cause a push on exit. That is the two tires are trying to follow a larger arc that is projected to somewhere off the race track. If we can overcome this tendency, then the rear tires will fight each other and there will be loss of grip.

More Stagger - Excess stagger causes the tires to want to follow a smaller radius arc and because of the centrifugal force cannot force the front to follow that arc. Deficient stagger will usually show up as a loose condition because the rear tires are turning at different speeds, and one or both rear tires will slip causing loss of traction.

Important Considerations - So, we can see that correct stagger is a must. To arrive at the best tire sizes, we need to consider: 1) what is the correct stagger amount for the track we are running, 2) how much might each tire grow and how soon, and 3) How can I quickly solve a stagger problem using the four tires on the car.

Correct Stagger – This is the amount of stagger that will produce equal revolutions of the rear wheels when driving around the turns. For rear wheel drive cars, stagger is not as serious a consideration at the front of the car because the tires are not connected by an axle like the rear tires are.

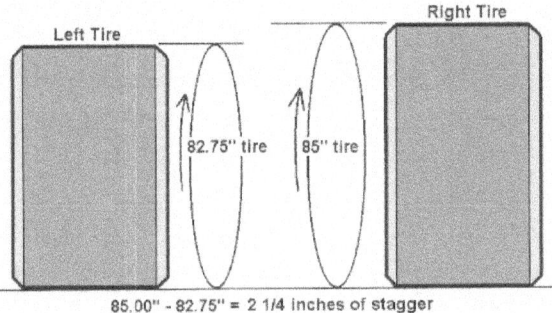

Stagger is the difference in circumference of two tires on the same end of the car. Take a measurement around each tire near the middle of the tread. Subtract the two measurements to find Stagger or "Roll Out" as it is sometimes called. Always inflate the tire to race pressures before measuring.

We can calculate what our stagger needs to be, or refer to charts that show the correct stagger for the radius of the turns, the track width of the two tires, and the average tire sizes.

Front Stagger - A definite consideration for front stagger is stability under braking. Because there is a lot of load transfer to the front under braking, the front tires and brakes work harder than the rears. If both wheels are turning the same number of revolutions, then the car will brake mostly in line with the direction the car is traveling.

If the Left Front wheel is turning faster (i.e. a smaller tire circumference and therefore more stagger) than required, the car may pull to the left on entry under hard braking. If the front stagger is too little, the car may not pivot as needed and develop a push on entry.

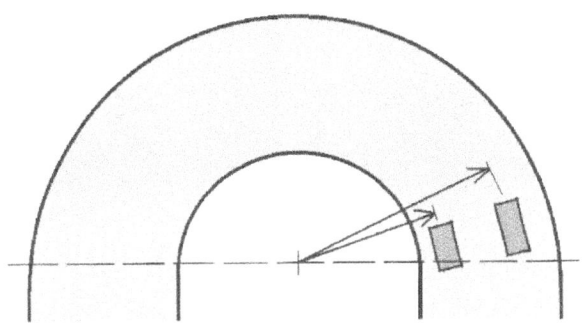

Radius of inside tire = 200 feet or 2400 inches (12" x 200')
Radius of outside tire = 2465 inches (2400 + 65" track width)
Outside tire travels 2465 x 2 x PI (3.1416) ÷ 2 = 7744 inches
Inside tire travels 2400 x 2 x PI ÷ 2 = 7540 inches
Outside tire = 85 inches in circumference
Inside Tire = 85 x (7540 ÷ 7744) = 82.75
Correct Stagger = 85 minus 82.75 or 2 1/4 inches.

This shows how to calculate the stagger the car needs in order for the inside and outside tires to rotate at the same speed and follow the arc of the turn. This is for a flat track and the track banking angle does affect how much stagger we will need. The banking reduces the stagger amount. To illustrate that, for a track banked 90 degrees, we would need zero stagger.

The general rule is to maintain a front stagger amount that is close to the correct stagger needed for equal revolutions of the wheels in the radius the wheels are running in. Remember that the radius on entry is much more than at mid-turn, so you will need less stagger than at the rear of the car.

Anticipating Tire Growth – Part of the "art" of tire stagger selection is being able to predict how much each tire will grow due to increases in pressure as the race progresses. As a rule, both of the right side tires (for a left turning car) will grow more than anticipated. But because tire growth is mostly influenced by pressure increase due to heat, it is possible that the left rear could grow more than the right rear tire if it is slipping and gaining excess heat.

Example: If we have a car with too little stagger in the rear, but one that turns well enough to overcome the tendency to push, then the left rear tire will slip and heat up. That is because, in most cases, the right tire is the more heavily loaded tire at mid-turn and is less likely to slip. One of the rear tires must slip, so it ends up being the left rear.

Turn Radius	\multicolumn{9}{c}{Right Rear Tire Circumference}
	80"	82"	84"	86"	88"	90"	92"	94"	96"
100'	4.24	4.34	4.45	4.56	4.66	4.77	4.87	4.98	5.09
125'	3.39	3.48	3.56	3.65	3.73	3.81	3.90	3.98	4.07
150'	2.83	2.90	2.97	3.04	3.11	3.18	3.25	3.32	3.39
175'	2.42	2.48	2.54	2.60	2.66	2.72	2.79	2.85	2.91
200'	2.12	2.17	2.23	2.28	2.33	2.38	2.44	2.49	2.54
250'	1.70	1.74	1.78	1.82	1.87	1.91	1.95	1.99	2.03
375'	1.13	1.16	1.19	1.22	1.24	1.27	1.30	1.33	1.36
500'	0.85	0.87	0.89	0.91	0.93	0.95	0.97	1.00	1.02
1250' Daytona	0.34	0.35	0.36	0.36	0.37	0.38	0.39	0.40	0.41

This chart shows typical stagger amounts for a sample car with a 65 inch track width and running on a twelve degree banked circle track.

Handling Affects Stagger - Most of the time, tight or loose handling will determine which of the right side tires will grow the most. A severe push will heat up the right front tire and cause it to grow beyond normal. The rear tires may maintain correct stagger because neither of the rear tires is being abused.

A much more common problem is the tight / loose syndrome. That is when the car is somewhat tight, but with increased steering input, the car reverses handling balance at mid-turn and goes loose off the corner. The right rear tire suffers abuse as it slips and gains heat and pressure. Stagger will grow as a result.

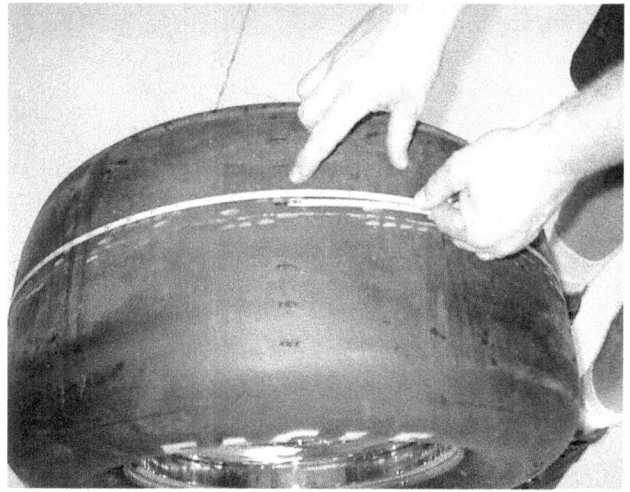

To find stagger the easy way, follow this simple method that I learned from a friend many years ago, and you will never need to subtract tire sizes again. Measure the larger tire first. Then remember the number.

Wrap the tape measure around the smaller tire and note where the large tire stagger number falls near the end of the tape. Read the measurement from the end and you have the stagger. You read the stagger directly instead of having to subtract each time.

Working With Stagger – Stagger is a very important part of race car setup for those cars that need it. Understanding what stagger is and what influences it helps when it comes time to work with a real race car. But understand that every car is different and every track has its own characteristics, so finding the right stagger is mostly a product of time and experience.

Exam - In The Context Of This Lesson:

Tire Stagger Is?

1) When one tire is ahead of the other at one end of the car

2) The inside tire having a larger circumference than the outside tire

3) The inside tire having a smaller circumference than the outside tire

4) Not important for cars that turn the same direction

Tire Stagger Is Measured How?

1) By knowing the width of the tires

2) Measuring the track width, center to center of the tires

3) Knowing the radius of the track

4) Finding the difference in circumference of the two tires on one end of the car

Excess Stagger Results In?

1) Loss of Grip

2) A car that may be loose into the corner

3) A car that is loose off the corner

4) All of the above

Too Little Stagger Results In?

1) A push off the corners

2) A car that is loose through the middle of the turns

3) A hotter inside tire temperature

4) All of the above

Correct Stagger Is A Product Of Knowing?

1) The radius of the turns

2) The track width of the tires on the same end of the car

3) Average tire sizes

4) All of the above

Lesson Thirty – How Brakes Work

Glossary:

Brake – A system in a race car that slows and stops the car. Most every system in todays racing is a disc and caliper system.

Brake Fluid – A hydraulic fluid used in brake systems that is usually glycol-ether, mineral oil or silicone based. Brake fluid is not compressible.

Brake Disc – A disc of various diameters, thicknesses and designs that is mounted to the wheel hub and is grabbed by the brake caliper to slow the motion of the wheel and therefore slow the race car.

Brake Caliper - In a disc brake system, a clamping device that attaches to the spindle for a double A-arm suspension or the axle tube for a straight axle rear end. Fluid is forced by the brake master cylinders into the caliper to push pistons against the brake pads and the brake rotor to slow the car.

Brake Pads – In a disc brake system, a pad that is positioned between the caliper piston and the brake disc to apply friction to the disc to slow the car.

Brake Bias – The tuning of the amount of stopping force of the front and rear brakes. There are various ways to create this bias.

Cooling Vanes – Air slots cast within the rotor to direct air in and through the rotor to cool it. Air comes in at the center of the rotor and exits at the outer edge by the use of centrifugal force as the rotor rotates.

Over View – Speed in a race car is not only about accelerating off the turns and getting through the turns, but involves deceleration when going into a turn. The longer it takes to decelerate, the longer the lap times and the less performance the car has.

Brake systems for race cars must be designed for the type of racing they will operate within. A system designed for high heat and severe applications won't work very well in uses where light braking takes place over shorter periods of time. We'll explain all of that.

Just know that there are certain undisputed characteristics of race car braking systems that you must know. First let's describe a modern race car braking system.

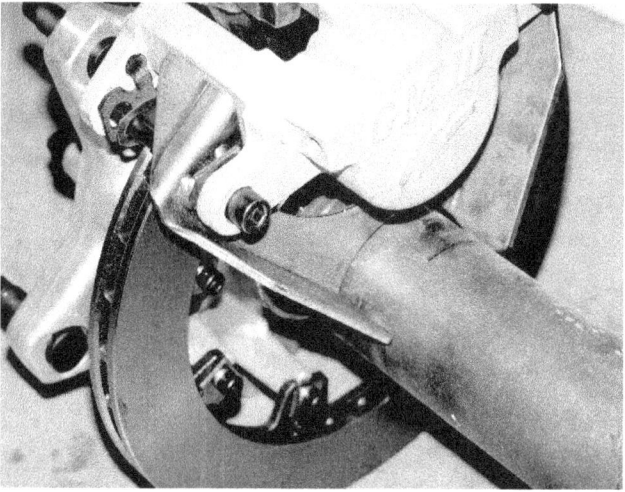

Most modern brake systems are disc systems. A disc is clamped by a caliper that holds brake pads. The pads have a friction property that grabs the disc slowing the car. The size of the caliper, the makeup of the pads, and the sizes of the master cylinders all help to create and determine the braking force.

Brake Components – Most modern brake systems for race cars are of the disc design. Most modern road cars use the very same systems with variations on the size and properties of the components. We first have a brake pedal that the driver pushes on when they want to slow the car. This pedal then pushes on a master cylinder mounted somewhere behind the brake pedal.

The master cylinder applies a force to non-compressible brake fluid through a brake line (tube that the fluid can flow through) into a brake caliper near the wheel.

The caliper has one or more pistons that are forced out against a set of brake pads. These pads are made of a material that has friction and rubs against a brake rotor, or disc that is attached to and rotates along with the wheel.

It is the clamping force on the disc that slows the car. The greater the force applied to the fluid, the greater the clamping force. The greater the friction the pads have, the more grip they will have on the disc and the greater the stopping force.

How Much Braking Force – We can either have too much or too little braking force for the weight of the race car and other factors such as tire grip, loading on the tires, track surface grip, etc.

For dirt tracks for example, if we use too much braking force, the tires will just slide across the track and not slow the car. This dirt surface provides very little grip for the tires. So, a lighter touch on the brake pedal and/or less grip in the brake system helps the tires to maintain grip when we brake.

For high grip surfaces like asphalt, if we have good tires, and aero downforce adding a lot of load to the tires, we can brake very hard and slow the car much faster. This is how it works with Formula One cars. They can decelerate from near 200MPH to 50MPH in just a few seconds.

Brake Master Cylinders – As the driver pushes on the brake pedal in the cockpit, a rod connected to a master cylinder is pushed and that pushes on a piston. This piston in-turn pushes on the brake fluid which is an oil based fluid that will not compress. This fluid passes through brake lines to the brake calipers.

Most race car braking systems use two brake master cylinders, one for the front brakes and one for the rear brakes. In this way, we can tune the braking force between the two ends of the car. More on that later.

Brake Calipers – As the brake fluid is pushed into the caliper, it applies a force to the pistons within the caliper. These pistons then push on the back of the brake pads to apply a force to the brake rotor to stop the car.

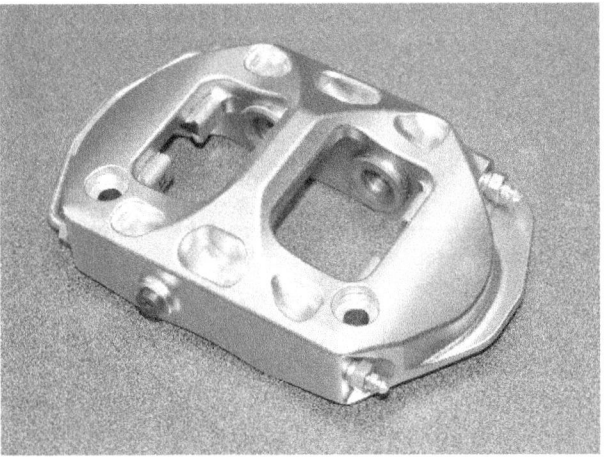

Modern race car calipers are one-piece designs that incorporate two pistons, or more, on each side to press two brake pads against both sides of the brake rotor.

Once the pedal is released by the driver, the fluid moves back into the brake lines and into the master cylinder waiting for the next application.

Brake Pads – Companies that make and sell brake parts have a wide range of brake pads available. The ones a particular race teams needs will depend on how they are to be used. Considerations include: 1) how much braking will take place for what period of time?, 2) What speeds are we braking from and to?, 3) How much grip does the track surface provide?

A brake pad that has a lot of grip might not survive high heat and repeated and prolonged use because we can think of them as softer. The material grips better because it sticks to the rotor better. Think of rubber and the grip it has verses steel.

If we put load on something that is coated with rubber and try to drag it across the floor, it will be very hard. If that same load were put onto something lined with steel, it would move much easier. In the same sense, the rubber would wear out quickly whereas the steel would last a long time.

In race car braking, the above example relates to the brake pads. We might want better braking force from a softer pad, but if it wears out before the end of the race, we end up with no brakes.

So, the choice of brake pads is a compromise between having more grip and wanting the pads to last. There is a happy medium for every type of race car.

Brake Rotors – The brake rotor is attached to the wheel and rotates with the wheel. To stop the car, the caliper with the brake pads must grab the rotor and apply friction that will stop it from spinning. That is how brakes work.

There are different thicknesses of rotors, different diameters for different uses, and different overall designs for the face (sides) of the rotor and the cooling vanes (slots within the rotor that route air to cool the rotor).

The brake rotor is attached to the wheel hub and rotates with the wheel. Note the cooling veins in the middle of the rotor. In this design, they are angled to the rear at the top. This left front disc is cooled by forcing air from the center of the rotor to the outer edge through the use of centrifugal force.

Rotors can be different diameters and the larger the diameter, the more efficient the braking system. That means, the caliper and pads can apply more stopping force to a larger diameter rotor. It is a product of leverage. The rotor size versus the tire diameter dictates the leverage force.

What is applying the force is the wheel and tire. The lever is the radius of the tire. It's length compared to the radius of the rotor creates a motion ratio. The larger the ratio, the more force the tire puts on the brake rotor. When the rotor is larger, the motion ratio is smaller and the tire has less leverage to counter the braking force and we stop faster.

Which Rotor To Use - We use thinner rotors for applications where we use less brakes. An example would be a car racing on a very fast track at a fairly constant speed like at Daytona. Even at Indianapolis or maybe Charlotte, where the cars must slow to go through the turns, the drivers use engine braking and not the actual brakes.

Thicker brake rotors are used for short track applications and road racing where there is constant and heavy braking that cause the rotors to experience a lot of heat and wear. It is in this application where we spend most of our time evaluating and designing brake systems.

Braking Force – The amount of braking force we will have in our braking system is dependent on the following factors: 1) amount of pressure applied to the brake pedal, 2) brake pedal leverage to the master cylinders, 3) the surface area of the master cylinder piston face, 4) the total surface area of the brake caliper pistons, 5) the brake pad surface area, 6) the friction properties of the brake pad compound, 7)the rotor diameter size, i.e. leverage ratio, 8) rotor surface friction, and 9) the heat generated by the system when in use.

The amount of braking force a driver can use is directly related to how much grip the tires have. If the driver over-brakes, then the tires will slide. The driver must apply as much braking force as is possible without causing the tires to lose traction.

The cutaway of a brake master cylinder shows the piston. The shaft to the right connects to the brake pedal and the brake line to the calipers comes out of the left end in front of the piston. The spring helps the piston return to the resting position after the braking is completed and the force is off the brake pedal. The brake fluid reservoir is on top over the piston and bore.

For race cars with high aero downforce, the braking force can be much higher in the initial phase of braking where the speeds and aero downforce is the greatest. As the car slows, the aero downforce becomes less and the loading on the tires also becomes less. Then, later in the turn entry when the tire loads are less, we must reduce the braking force to prevent the tires from losing their grip on the racing surface.

A race car transfers weight when it is decelerating, as when braking, and so the front tires will have more loading and grip than the rear tires in most situations. Therefore, we can use more braking force on the front tires than on the rear tires.

If we used the same amount of force for both the front and rear tires, one of two things would happen, either the front tires would not be braking to their maximum potential, or the rear tires would lose traction and slide.

If we design our braking system so that the front brakes will generate more braking force than the rear brakes, we can then maximize our braking system.

Brake Bias Explained – Because we cannot, in most cases, apply equal braking force to the front and rear tires, we must create brake bias, or a differential in braking force for the front verses the rear systems. The front brakes will have more force than the rear brakes in almost every case.

Brake bias can also be applied between opposing brakes on the same axle, or end of the car. On dirt cars, the designers recognize that the right front tire must always maintain grip with the track surface in order to turn the car, so they sometimes either reduce the braking force to the right front brake, or lock it out and eliminate any braking force to that corner of the car.

In a typical installation for most race cars, the three master cylinders are shown. The closest is the clutch master. The other two are the brake master cylinders, one for the front and one for the rear brakes. We can install different master cylinder bore and piston sizes for each master to create brake bias for the front and rear brakes.

This differential in braking force could also be applied to a road racing, or formula car where there are high G-forces in high speed corners that generate a lot of load transfer off of the inside tire. That tire cannot generate as much braking force as the outside tire without losing it grip on the racing surface.

If the designers could create a differential braking system that reduced the braking force to the inside wheel, then the driver could brake harder without causing brake lockup of the inside wheel.

How To Create Bias - There are several different ways to create brake bias. Since we want to apply as much braking force as possible, we can divide the force applied by the driver between the front and rear brakes.

Brake master cylinders come in different sizes as to the diameter of the pistons and the bore size associated with the piston size. The smaller the piston, the greater the force on the fluid for the same pedal pressure applied by the driver. This is a very important concept to understand.

Pressure in our braking system is measured in pounds (or any other measure of weight) per square inch. If we apply 50 pounds to a one square inch piston, we have applied 50 pounds per square inch of force to the calipers.

If we double the size of the master cylinder piston to two square inches, we will be providing only 25 pounds per square inch of force to the calipers. Again, the smaller the size of the piston in the master cylinder, the greater the force on the calipers.

So, we can use a larger diameter master cylinder for the rear brakes than we use for the front brakes to get the brake bias we need that will provide greater braking force to the front verses the rear brakes.

Another way to create brake bias is to use different sized brake calipers. The piston sizes and numbers in the calipers dictate how much brake line force will be applied to the brake pads. The force verses caliper piston size is the inverse of how the master cylinders work.

The larger the piston in the caliper, the greater the force on the brake pads will be. If our line pressure were 50 psi (pounds per square inch) and our caliper piston areas was 2.0 square inches, we would be applying 100 pounds of force to the brake pads.

If the caliper piston area were 3.0 square inches, we would be applying 3.0 times 50psi = 150 pounds of force to the brake pads. If we install larger calipers in the front than at the rear, we can create brake bias using the calipers.

When we install the brake calipers, we need to make sure that they are positioned correctly in height and that they are parallel to the disc rotor. Clamping the caliper to the rotor before welding the mounts ensures a perfect fit.

Brake Bias Balance Bar - Once we have established a good ratio of braking force for the front to rear brake bias, we can use what is called a brake bias balance bar that is attached to the brake pedal. This bar can be adjusted by the driver to apply a bias of force between the front and rear master cylinders.

This device is a fine-tuning tool only and should not be used to create the initial brake bias we need due to tire loading differentials caused by load transfer on braking.

Brake Fluid – One thing that is important to know about brake fluid is that it can absorb moisture from the air. Any water that finds its way into the fluid will expand and boil when the fluid heats up causing partial or complete brake failure.

Heat is a bi-product of braking. The primary cause of the heat is the brake rotor rubbing against the brake pad and heating to many hundreds or thousands of degrees. The rotor transfers that heat on to the fluid through the brake pads and calipers.

Race teams need to regularly change the brake fluid in the system and add fresh, non-contaminated fluid that does not contain water. Other regular maintenance includes inspection and changing out of the O-rings that seal the master cylinder pistons and the caliper pistons.

Exam - In The Context Of This Lesson:

The Amount of Braking Force We Can Apply Is Regulated By?

1) The grip of the tires
2) The frictional properties of the racing surface
3) The loading on the tires
4) The force applied to the pedal
5) All of the above

The Master Cylinder Provides More Force When?

1) It is larger than the rear master cylinder
2) It has a smaller piston
3) The brake fluid lines are larger
4) It has a larger piston

The Caliper Provides More Force When?

1) It is smaller than the rear master cylinder
2) It has smaller pistons
3) The brake fluid lines are larger
4) It has larger pistons

Brake Rotors Have More Stopping Power When?

1) They are smaller in relation to the tire diameter
2) The disc is thicker
3) The disc is thinner
4) They are larger in relation to the tire diameter

Why Do We Need Brake Bias?

1) Because of load transfer when accelerating
2) To keep the brake force equal on all four tires
3) To provide greater braking force on the rear tires
4) To compensate for unequal loading on the tires

Brake Bias Can Be Created By?

1) Using different sized brake calipers
2) Using different sized brake master cylinders
3) Using brake pads with different compounds
4) Using a bias balance bar
5) All of the above

Lesson Thirty-One – Summation RCT Level One

Congratulations, you have completed the Race Car Technology course, Level One. This final Lesson will outline all that we have accomplished in this course. In Race Car Technology Level One we have introduced you to all of the parts and pieces of the race car and talked about what each ones role is with the car.

We first outlined all of the goals of Level One. We proceeded to go through all of the systems on a race car and help you understand how each one works.

We first taught you about the Goals in setting up and racing a car and what we wanted to achieve in Level One.

We then told you what makes a race car turn. This is very important to understand because all we do is centered on first making the car turn well.

We described the various geometry part and pieces and how they might interconnect.

We discussed the steering systems and how things like Ackermann can affect the way the car turns.

We told you about Alignment and how this works with the other setup parameters to make the car perform at it peak.

We went through all of the Anti's in a race car and described how that works. We talked about springs, motion ratios, and the influence of springs.

Then we told you how shocks control the spring rates to control chassis motion and weight distribution. We described all of the parts of a modern racing shock and how they control movement of the shock, and in conjunction with that, the motion of the chassis.

You learned about racing tires, how pressures, camber, and temperatures affect the tires. Everything we do in race car engineering is intent on making the tires work as hard as possible under the existing conditions.

And finally, we learned about racing brake systems. The faster you can make the car slow, the faster you can go.

Level Two goes on to explain and teach how to work with all of these parts and pieces. Each of the suspension parts will need to be worked with and set so that their contribution helps make the car faster and more consistent.

About the Author

Bob Bolles has been a hands-on motorsports engineer for over twenty-five years. Although he holds a B.S degree in the Mechanical Engineering, his skill and experience with racecars comes from working directly on the chassis and with many hundreds of race teams. He likes to think that he comes from much the same mold as a few others before him such as his friend, the late Dr. Smokey Yunick, who also wasn't afraid to "get his hands dirty" to effect change. Like Smokey, Bob loves this sport and enjoys the interaction with others who also love it.

The nice thing about racing is that you cannot BS what you know or what you develop. It either works or it doesn't. Proof is just a few laps away. Before Bob started his research into racecar dynamics and engineering, he struggled with the very same problems that most race teams continue to struggle with today. He just knew there was a better way and if he just looked hard enough, the answers might come. Well, they did. That information is shared in the pages of the RCT series.

Over the years, Bob has worked with a high degree of success on virtually every type of stock car raced in the United States, as well as sophisticated formula type race cars. Whether it is an asphalt stock car, modified, or a dirt late model, his techniques and methods have improved handling and opened the door for winning. He has engineered cars that have won major asphalt late model championships, touring championship, modified championships, road racing championships, and dirt late model races including the Dream and the World 100 at Eldora Speedway.

It is this broad level of experience and his development of new technologies that qualifies him to write about these important subjects. Not a day goes by that Bob isn't speaking with or helping racers. With contacts all across America, he is truly in tune with the pulse of auto racing on a technological level that very few individuals enjoy.

The software Bob developed in the mid-1990's is still being used by championship winning teams throughout the U.S., Canada, Australia, New Zealand and Europe. His dream of being able to help all racers in their pursuit of success and enjoyment is fast becoming a reality.

Bob has been a professional motorsports technical writer serving as the Senior Technical Editor for Circle Track magazine for over fifteen years and currently as a technical contributor to Speedway Illustrated magazine. He is the author of Stock Car Setup Secrets and Advanced Race Car Chassis Technology published by HP Books.

Race Car Technology- Level One represents Bob's first book in this series. Bob is also the author of RCT - Level Two and RCT – Level Three. These books are being used by college instructors to educate future motorsports engineers.